MW01378135

A FEW OF MY FAVORITE THINGS

Birds, Birds, Birds, Birds, Birds

ROLAND H. WAUER

Author's Tranquility Press
ATLANTA, GEORGIA

Roland H. Wauer/Author's Tranquility Press
3800 Camp Creek Pkwy SW Bldg. 1400-116 #1255
Atlanta, GA 30331, USA
www.authorstranquilitypress.com

Ordering Information:
Quantity sales. Special discounts are available on quantity purchases by corporations, associations, and others. For details, contact the "Special Sales Department" at the address above.

Favorite Things/Roland H. Wauer
Hardback: 978-1-965463-12-3
Paperback: 978-1-962859-27-1
eBook: 978-1-962859-28-8

INTRODUCTION

A Few of My Favorite Things reflects my many years of birding. As a hobby, actually an obsession, birds have dominated a substantial part of my life. In support of my birding activities, I have been fortunate to have worked/lived in several national park areas and/or National Park Service offices. In succeeding order, they included Crater Lake in Oregon; Pinnacles and Death Valley in California, Zion in Utah, Big Bend in Texas, Santa Fe in New Mexico, Washington D.C., and Great Smoky Mountains in Tennessee and North Carolina. Most of the birds included in A Few of My Favorite Things, is the result of the diverse opportunities provided.

A Few of My Favorite Things includes a grand total of 90 bird species, 84 of which are illustrated with photographs. Additional photographs are included to illustrate pertinent habitats and/or characteristics.

Sunrise! The start of a new day! Rejoice! I have enjoyed some amazing dawns! Watching the eastern sky as it begins to brighten, I anticipate what the new day might offer. Some days the eastern sky is a deep reddish cast that lasts too short a time before it brightens and fades. Other days lack the deep reddish glow but still gradually brightens my surroundings. I then look about me and wonder what that new day will provide. Each is special!

I recall some truly special sunrises when the sky literally become filled with tremendous numbers of waterfowl arising from their overnight roosts. I remember a dawn at Mad Island Marsh when thousands of snow geese arose up from the fields where they had spent the night, and began their search for another field where they could feed and socialize. Christmas bird counts provided those opportunities, and they have long provided special memories. Rising before dawn, locating a position where I could see that great lift of birds, was truly mesmerizing. There's no place more empowering than to be than in the present moment.

Waterfowl at dawn

I can't number the many times I have awakened to the dawn chorus of birds. Camping out, whether that be in deserts or tropical forests, all begin with amazing bird sounds. And those dawn songs are different than those heard the remainder of the day.

I recall listening to **Curve-billed Thrashers** one early morning at Organ Pipe Cactus National Park in Arizona. I wrote about that morning's birdsongs in *Songbirds of The West, Personal Encounters*, thusly:

> If the morning songs of Cactus Wrens does not provide an adequate wake-up, surely the "snap-song" of the Curve-billed Thrasher will. Larger than Cactus Wrens, Curve-bills also commonly sit at the top of saguaros at the start of each day. They warm up with loud, almost explosive "whit-wheet" calls but before long shift into a more elaborate and melodic song that sounds very much like that of its cousin the Northern Mockingbird. One February morning in the campground I timed several Curve-billed Thrasher songs that lasted eight to a full thirty seconds each.

Curved-billed Thrasher

I recall another dawn on Gulf Islands National Seashore, where **Brown Thrashers** dominated Davis Bayou Campground. Brown Thrashers literally were everywhere. On one brief stroll around the campground, I counted 24 individuals, from the very treetops to almost every shady patch along the edge. In full sunlight, Brown Thrashers were a gorgeous, bright rufus color. They all possessed long legs which are utilized for running as well as hoping, scratching, and other useful functions. One of those activities provided thrashers with their name: they actively thrash the debris aside when feeding. But their most outstanding characteristic is their singing ability.

I had long thought the Mockingbird was the supreme songster in the bird world. Mockers can imitate almost any bird, not to mention house cats and a lot of other creatures. But Brown Thrasher songs possess a deeper and richer quality, and they can be just as loud and obnoxious. Kent Rylander also expressed his appreciation of Brown Thrasher songs when he described its song and call in *The Behavior of Texas Birds:*

> Song: somewhat like the Northern Mockingbird's but throaty and with fewer appropriated sounds. Compared with the mockingbird, the phrases are generally more abrupt and in twos instead of threes or more; they sound to some listeners like a telephone conversation: *Hello, hello, yes, yes, who is this? Who is this? I should say. I should say, how's that? how's that?* Call: hisses, clicks, and whistles, including a loud *spack.*

The **Long-billed Thrasher** is one of the "Texas specialty" birds; it is found nowhere else except in South Texas and southward through eastern Mexico to southern Veracruz. In Texas, it resides in dense bottomland thickets. Unlike Curve-billed Thrashers, which are an overall tan color, long-bills are darker with several lines of black streaks on their otherwise white underparts. And they, too, sing a varied song that is much like that of Brown Thrashers.

Long-bill. Thrasher

Gary Clark, in *Book of Texas Birds*, wrote that the long-bill's voice is a "whistled *cleeooeep* sound. Song is loud with phrases repeated. Sounds like *tsuck, kleok*. Not a mimic like other thrashers." In *Guide to the Birds of Mexico and Northern Central America*, Steve Howell and Sophia Webb's interpretation is that: "song a loud, rich to slightly scratchy warbling, phrases often repeated 2-4x."

My encounters with this rather secretive thrasher have primarily been in Tamaulipas scrub habitat, locally known as "South Texas Brushlands," and including thorn scrub and mesquite grasslands. This region generally runs from Corpus Christie and Hazel Bazemore County Park southward. In reviewing *Birding Texas*, I listed Tamaulipas scrub at Kingsville, Cayo del Gruillo, Park Chalk Bluff, San Antonio South, Chaparral Wildlife Management Area, Choke Canyon State Park, Lake Corpus Christi State Park, Lake Casa Blanca State Park, Laguna Atascosa National Wildlife Management Area, Brownsville, Santa Ana National Wildlife Refuge, Bentsen-Rio Grande Valley State Park, and Falcon State Park and area. Tamaulipas scrub covers a significant portion of South Texas.

I recall one occasion at Santa Ana, while walking a brushy trail I discovered a Long-billed Thrasher very nearby. It seemed to move with me although remaining well hidden in the tangle of shrubs and vines. I was reminded of Tim Brush's comments in *Nesting Birds of the Tropical Frontier*:

> The *mante* (dense thorn forest and thorn scrub) is classic Long-billed Thrasher habitat. When the canopy is 10-15 feet high, the understory may be fairly open, with stands of Texas ebony, coma, or Texas persimmon, but in other cases only a rabbit or thrasher would seem able to squeeze through the thorny brush. Thrashers are found in lower densities in tall riparian forests, mainly in areas with some understory.

Although all the members of the thrasher family (Mimidae) sing unique and distinct songs, the mockingbird being the best example. But these extremely vocal birds take second place to songs of thrushes, birds of the shadows.

Thrushes, members of the Turdidae family, are found worldwide. They are small to medium-sized ground-loving birds that feed on insects, other invertebrates, and fruit. One of their most memorable features is their lovely songs, often sang throughout the day during the breeding season.

The loveliest birdsong I know is that of the **Wood Thrush**. It is a flute-like "ee-o-ly" which is repeated over and over. Its song is never in a hurry, taking its time to make sure that each song is better than the one before. It has been described as a hymn of praise rising pure and clear from a thankful heart, 'ee-o-lay, ee-o-lay.'

There is something haunting about the song of the Wood Thrush. It is a sound of the deep, undisturbed forest, where few humans venture. Its incredible songs cannot soon be forgotten. Louis Halle wrote:

> The ease and leisure of the wood thrush's song is one of its characteristics. The singer is never shaken with effort like a house wren. Usually he sits motionless on a branch, at rest. Every few seconds (with the regularity of some marvelous mechanical toy) he lifts his head, opens his bill, and delivers himself of a brief phrase; subsiding then until another phrase has formed and is ready to well up within him.

Living in Springdale, Virginia, while working in Washington, D.C., I was able to run most evening within my neighborhood. I looked forward to every one of my runs, not just to shed the concerns of the work-day, but also to enjoy the lovely, relaxing songs of Wood Thrushes that were nesting in the adjacent woods.

I also recall the Wood Thrushes that I encountered at Chattahoochee River in Georgia. I later described the bird as a reasonably large thrush with a rufous back, head, and tail and white underparts with large black spots. It often feeds on the ground like its first cousin, the American Robin, and it also feeds on berries from fruiting shrubs in season.

Years later, when Betty and I were visiting every national park in the U.S. and Canada, my memory of Mammoth Cave in Kentucky was not the magnificent cave we walked through, but of the Wood Thrushes encountered in the adjacent forest. Wood Thrushes were in full song as we walked down the path that summer morning, beyond the park visitor center, past the hotel to the historic cave entrance, and onward to the floodplain of the Green River.

I counted at least a dozen individuals along the way. Their songs seemed even more vibrant as we descended the slope. Their only competition that morning came from busy Carolina Wrens, another reddish-colored songster of the forest. Next in abundance perhaps was the White-breasted Nuthatch. Their continual nasal honks provided communications between young birds and adults. I remember being totally relaxed that morning while enjoying the marvelous sounds of the forest. It seemed to me that it was the Wood Thrush songs that set the mood.

I have a long history with **American Robins**. It began when I was a youngster, about 14 years old, when, while shooting my bow and arrows, I killed a distant robin. And when retrieving my arrow, which had pierced that Robin, I suddenly felt terrible guilty for what I had done; in a sense, that incident started a life-long love of birds.

Years later, perhaps because of that earlier incident, I wrote *The American Robin*, that contains much of what is known about that marvelous bird. I included the following paragraph on the Robin's voice.

American Robins sing loudest and sweetest at dawn and dusk, especially in spring and early summer when nesting. Throughout their range, the Robin's song is usually the first birdsong heard in the mornings and the last song heard in the evenings. Aldo Leopold used a sensitive photometer to determine the light intensity during the morning songs of 20 songbirds near Madison, Wisconsin. He discovered that the American Robin begins its morning songs with as little as 0.023 foot-candle of light at 3:15 a.m. on clear April mornings. He also confirmed that cloudy conditions delayed morning songs and that bright moonlight at night enticed birds to sing. The American Robins also sing throughout the summer months and into September or October, when most other birds are silent.

American Robin

The interpretation or phonics of Robin songs varies with the listener. To my ear, the typical Robin sings "Cheer-up, cheerily, cheer-up, cheer-up, cheerily." Other writers have interpreted Robin songs somewhat differently. One interpretation stated that "Cheerily, cheery is a favorite rendering of this song, aptly suggesting by sound and meaning the joyous tenor of the phases, amid the liquid quality of the notes." No other bird sings for so many hours of the day and for so long a period of time.

In my closing of *The American Robin*, I wrote:

The American Robin is unquestionably North America's most widespread songbird, present in the far north during the summer months, throughout most of the United States year-round, and common in the southern parts of its range at least in winter. The result of this continental distribution is that the American Robin is our most familiar bird; it is the official state bird for Connecticut, Michigan, and Wisconsin; it is featured on Canada's two-dollar bills; and every child learns to recognize it almost before walking.

The Robin's familiar features – its robin-red breast, upright posture, and loud, ringing songs – are significant parts of our earliest outdoor memories. No other bird sings so many hours of the day and for long a period of time. Its size, breast color, robin's-egg blue eggs, and flutelike caroling are standard-settings in the bird world.

Robins also may be our most adaptable songbird, seeking out new breeding grounds whenever possible and adjusting to an amazing diversity of food, from a wide variety of invertebrates, such as earthworms, to fruits of many kinds. Robins use our lawns and gardens during the nesting season, our berry trees, shrubs, and vines in summer, fall, and winter, and our birdbaths whenever possible. They are one of our most courageous songbirds, defending their nests and nestlings against a host of threats.

These highlights conclusively prove that the American Robin is truly our most visible and beloved songbird.

Robins at birdbath

One additional feature about Robins is their love for water, as my photo above illustrates. I wrote about a related incident in *The American Robin*:

> I remember one early spring day along the Naval Oaks Trail in Florida's Gulf Islands National Seashore. The huge live oaks were filled with birdsongs, each species expressing their zeal for the coming season. Then ahead of me, somewhere to the left of the trail, I began to detect a strange, melodious hum. It took several minutes to reach a point on the trail where I was able to pinpoint the general location of the sound. By this time I began to see dozens of American Robins all about me, perched among the oaks, moving about from one spot to another, and arriving single or in flocks from elsewhere.
>
> I began to zero in on the principal source of the hum, soon recognizing that much of the harmony was supplied by mellow chips and partial songs of Robins. I left the trail and slowly worked my way through the woods to where I could see a shallow pond just ahead. By now the sound was considerably louder, and I could also distinguish hundreds of minute splashings. A few feet closer and I was able to peer through the undergrowth to across the pond. It wasn't until then that I understood the true cause of the hum. Hundreds of American Robins lined the shore or were perched on adjacent shrubs and trees. Those along the shore were bathing, dipping into the water and flipping it over their backs. They were spaced out shoulder to shoulder for 100 feet or more. Each bird seemed in pure delight! After a few dips and splashes, the bather would fly up to a low branch to shake and preen and chirp a few apparent notes of contentment. Its place on the shore was immediately taken by a waiting bird.
>
> In watching the estimated thousand or more bathers that morning, I was struck by their good manners and patience in waiting their turn to bathe. Although I noticed some posturing during a bath, or an occasional bill jab or gaping when the next bather got too close, the entire flock reminded me of lines of shoppers streaming to the checkout and waiting their turn. It was a marvelous experience!

All the thrushes, which include Robins, possess marvelous songs. But perhaps my favorite is that of the **Varied Thrush**. I included the Varied Thrush's amazing song in my chapter on Mount Rainier in *Birds of the Northwestern National Parks*:

The eerie, bell-like song of Varied Thrushes echoed across the meadow. A second later I had located this robin-sized thrush at the very top of an old snag. With binoculars I could see the bold black band across its deep-orange-colored chest, black cheeks with contrasting orange eyelines and throat, and dark gray back and wings with orange bars. It sang a dozen or more phases and then was silent.

Each phrase of its song had begun on a slightly different level, but each was delivered with enthusiasm. Some birders have described its song as a long-drown-out quavering note that some compare to the quality of escaping steam; after a short interval the note is repeated in a higher pitch, then again in a lower, "ee-ee-e-e-ee." Varied Thrush songs have a meditative quality due to their deliberation and all the strangeness due partly to their quality and partly to the complete invincibility of the singer. Roger Tory Peterson, in *A Field Guide to Western Birds*, described its song as a "long, eerie, quavering whistled note, followed, after a pause, by one on a lower or higher pitch." I have imitated this song by whistling it at the highest possible pitch.

Mount Rainier

In Birds of the Northwestern National Parks, I wrote:

> The Varied Thrush is unquestionably one of the park's most charismatic songbirds. Its eerie songs often echo through the forest from hidden places, giving the listener the feeling either of mystery or enchantment, depending upon one's perspective. Slightly larger and, at first sighting, seemingly like the American Robin, it possesses very different plumage: orange throat and eye stripe against a black cap and cheeks, and blackish back and wings with bold orange wing bars. Its reddish-orange chest is crossed by a bold black band. And like the Robin, during the breeding season it sings from the first light of day until dusk.

The **Hermit Thrush** usually stays in the upper canopy and requires patience to see it well. It possesses a rather distinct but somewhat varied song. According to my hearing it sings "che che chechezee de-de" or "we we we wezeeee che-che," and sometimes even adds an additional "che-che." In each case, the first three slightly ascending notes are followed by a buzzing note and two descending notes.

High country hikers are most likely to experience its incredible music. Naturalist John Terres provides us with the best description: it "opens with clear flutelike note, followed by ethereal, bell-like tones, ascending and descending in no fixed order, rising until reach dizzying heights and notes fade away in a silvery tinkle."

The Hermit Thrush is a bird of the conifers. So, finding this thrush at ground level is serendipitous as this bird stays in the shade and rarely spends any time in sunny places. When I have found one on the ground, instead of immediately flying away, it usually will freeze in place. Then is when I have seen its brown back, pale rusty tail, white eye rings, and spotted breast. Truly a lovely bird!

The breeding range of the Hermit Thrush is widespread in the northern states from Alaska to Maine, with southern extensions within the Rocky Mountains and the northern portion of the Appalachians. Wintertime populations extend across the southern states from central California eastward through Texas to south Florida and most of Mexico.

I first got to know this bird well while working at Zion National Park in southeastern Utah. In *Birds of Zion National Park and Vicinity*, I wrote that the Hermit Thrush is a "common summer resident; uncommon migrant and winter resident." There are numerous nesting records within the park's highlands, "post-nesting and/or fall migrants may appear in the lowlands as early as late August, there are a few scattered reports in September and October, and numbers increase considerably during November.

I banded more than a dozen individuals in Oak Creek Canyon from November 3 to December 11. Winter birds can usually be found on the Virgin River floodplain as well as in various side-canyons. By mid-April, however, spring migration is underway, and singing soon becomes prevalent in the highland forests.

Grand Canyon, North Rim View

I also wrote about the Hermit Thrush at Grand Canyon National Park in *The Visitor's Guide to the Birds of the Rocky Mountains National Parks*:

> In summer, the song of the Hermit Thrush, echoing from the conifer stands, can be one of the most appealing of all the highland birds. This bird is often heard but seldom seen, unless one visits the shadowy forest. Even then one must be quiet and patient and wait until it makes the first move. It often stays close to the forest floor, although it will sing its territorial song

from the very tips of the highest conifers. It may suddenly appear at the base of a tall conifer or among its heavy branches and sit like a sentinel until it is convinced you are not a threat.

North American Warblers, of the family Parulidae, are small migratory songbirds that travel long distances from as far as South America up to their breeding grounds as far north as Canada. They are active and usually brightly colored little birds that rush through nesting to leave for their wintering grounds soon afterwards. They leave us with a memory of yellow and green and missing their marvelous diversity of songs.

Warbler songs never have the flute-like quality of thrushes, but most possess a more complicated series of notes which are equally appealing. One of my favorite warbler songs is that of the **Townsend's Warbler**, a lovely bird of the American Northwest. It usually is found in the high foliage during the breeding season, although they commonly search for food in lower areas and often in the company of other songbirds. Males are a particularly colorful bird with a black throat, cap, and cheeks, bordered with bold yellow markings and yellow flanks, heavily streaked with black.

Townsend's Warblers sing a wonderful somewhat buzzy "jeer jeer jeer je-da," which rises gradually at first and then ascend sharply with the "je-da." Others have described its song as "weezy weezy weezy weezy seet" or "tsooka tsooka tsooka tsook tsee tsee!" No matter one's interpretation of its song, Townsend's Warblers are one of America's most colorful black and yellow songsters.

North Cascades N.P.

I recall a marvelous day at North Cascade National Park in Washington, sitting on the switchbacks of Cascade Pass, enjoying the early morning ambiance. The nearby montane forest was full of birdsongs. Townsend's Warblers were especially vocal that morning, singing an assortment of breezy songs. Several seemed to be singing a rendition that sounded most like "hea, yea, yea, sex-y."

The air that morning also was filled with other birds. I wrote about that occasion in *Birds of the Northwestern National Parks*:

> Overhead, Black and Vaux's Swifts zoomed here and there in their continuous search for insects. I wondered if the Black Swifts were nesting under any of the numerous waterfalls visible from the parking area. A pair of Vaux's Swifts, much smaller than the Blacks and with whitish underparts, spent considerable time near the top of a tall fir tree, and on one pass they disappeared altogether; they apparently were nesting in a high cavity. Several Pine Siskins and a pair of Red Crossbills passed overhead. Then a brightly colored, male Red-breasted Sapsucker flew out of the forest and disappeared into the alders; it had a bill full of insects, and I assumed it was feeding nestlings. It made two additional trips while I was watching.

While Townsend's Warblers usually are found in the high coniferous foliage, **MacGillivray's Warblers**, also a western species, stay low in the undergrowth, often along mountain streams. In Washington's North Cascades National Park, I was able to locate one of these birds among some alders by its very distinct songs, loud and clear "tswee, tswee, tswee, wit wit" or "swee swee swee swee swe."

I found one of the vocal males perched in the sunlight, as if were in a spotlight on a dark stage. Lighted as it was, its dark gray hood showed a deep blue cast, highlighting its white, broken eye rings, olive-brown back, and lemon-yellow underparts. What a truly lovely bird!

On another occasion, in Canada's Glacier National Park, I spend considerable time among the willows and alder thickets trying to coax a MacGillivray's Warbler into view. It stayed in the dense foliage at first, but I eventually was able to spish it close enough to see its gray and yellow plumage and white partial eye ring. A moment later it hopped onto a higher willow branch so that the sunlight highlighted its rich colors. I remember thinking that it had moved into the sunlight just for my benefit. I was most thankful!

Another warbler of the undergrowth and thickets is the **Yellow-breasted Chat**. The largest of our warblers, chats possess a bright yellow breast that gives them a very distinct appearance. And there are other characteristics that also set them apart from other warblers. I wrote about one encounter with a Chat at Dinosaur National Monument in Utah:

> A bright yellow-breasted bird suddenly arose from the riverside thicket, flew straight up 20 to 25 feet above the vegetation, and then, with wings flopping, tail pumping, and legs dangling, floated back to the same place from where it had first appeared. It sang a remarkable song during its descent, a jumble of whistles, clucks, mews, squawks, and gurgles. A moment later, it let loose with a series of "kuk kuk kuk" notes, very different from any it had produced in flight. I watched it through my binoculars as it preened itself, ruffled its feathers, and then began a completely new series of clucks, mews, yanks, and gurgles. Suddenly, it shot back into the air, and I watched a repeat of its earlier performance.

Yellow-breasted Chat

The Yellow-breasted Chat is a bird of the dense thickets that occurs along our rivers and streams. It can be loud and obnoxious in spring but quiet and secretive by summer and fall. Riverside campers in spring often find that this large, bright yellow warbler serves as an excellent alarm clock. While working one summer at Summer Home Park along California's Russian River, I taught a birding class to a half-dozen students. Although I was able to point out more than two dozen different species that day, the colorful Yellow-breasted Chat, common within the riverside vegetation, was their favorite.

Hooded Warblers also frequent low shrubbery. One afternoon at Jean Lafitte National Historical Park in Louisiana, while walking the Ring Levee Trail that skirts along some wetlands, I was attracted to a loud and clear song just ahead. Walking slowly toward the song, it sounded again, "weeta, weeta, wee-tee-o." A Hooded Warbler! Frank Chapman wrote that its song can be interpreted as "You must come to the woods, or you won't see me." To my ear, its song is lovely and sweet, but all too short; it seems to end just at its height.

Hooded Warbler

It took me several minutes to locate the singing migrant that day, and when I did, I found it walking about on the ground. I was then able to see it through some vegetation that apparently had at first screened it from me. My bird possessed a bright yellow face, coal black cap, collar, and throat, and a habit of flicking its white-spotted tail. Its contrasting plumage gave it a special, very appealing character.

I had a similar experience with a Hooded Warbler, probably nesting, at New River Gorge in West Virginia. I had camped at Babcock the night before and had awakened to a light rain at dawn. I left the campground and started down the Manns Creek Gorge Trail. Carolina Chickadees and Tufted Titmice greeted me first, and an American Robin glided out of the woods into the campground to search the mowed lawn for worms. Off to the right I detected the song of a Great Crested Flycatcher. And high above, an American Goldfinch was in route to some secret goldfinch location.

It was there, less than 200 yards into the forest, that the Hooded Warbler appeared. I stayed within a few yards of that site for the next hour; there was lots of bird activity. I first spent several minutes admiring that male Hooded Warbler until it departed as suddenly as it had appeared. Even though it flew off only a short distance, its “weeta weeta, wee-tee-o” continued. It seemed to have a ringing character to it, giving the entire woods its melodic presence.

I was then attracted to another warbler song, very close to where the Hooded Warbler had disappeared: “teacher teacher teacher.” The song was almost as loud as the Hooded Warbler’s but not so explosive. A second later I located the rather plain **Ovenbird**, sitting at eye level on a tree branch. The Ovenbird sang its teacher song again, four times in a row.

I made a low chip sound, and the Ovenbird immediately responded. It jumped onto a slightly higher branch and sang again. It was in better light now, and I could see its brownish-green back and head and its rather distinct whitish eye ring. It took me several seconds to see its orangish head stripe. Its overall brown-green coloration and striped breast provided this little warbler with marvelous coloration to help conceal it within the shadows of the forest. It then dropped back onto the forest floor and I watched it walk about in a jerky manner, searching for insects in every nook and cranny.

Great Smoky Mts., from Clingmans Dome

The Ovenbird is a reasonably common nester in the Great Smoky Mountains, residing within cove hardwood forest communities, generally typical of those special places found throughout most of the eastern deciduous forest. In my chapter on the Great Smoky Mountains, I provided the reader with what birds to expect within this community:

Common nesting species include the Yellow-billed Cuckoo; Eastern Screech-owl; Red-bellied, Downy, Hairy, and Pileated woodpeckers; Chimney Swift; Eastern Wood-pewee; Great Crested and Acadian Flycatchers; Blue Jay; Carolina Chickadee; Tufted Titmouse; White-breasted Nuthatch; Carolina Wren; Blue-gray Gnatcatcher; American Robin; Wood Thrush; Yellow-throated and Red-eyed Vireos; Northern Parula; Black-and-white, Worm-eating, and Hooded Warblers; American Redstart; Ovenbird; Scarlet Tanager; Northern Cardinal; Rose-breasted Grosbeak; and Brown-headed Cowbird. And in wet areas, the Louisiana Waterthrush, Kentucky Warbler, and Song Sparrow can usually be found.

During the six years I worked in Washington, D.C., I discovered that one of the ways I could get away from the bureaucratic rat-race was to spend time in the field watching birds. With the help of other birders, I gradually located several super birding sites. One of those was the towpath along the C & O Canal. The towpath was high enough above the waterway and vegetation that allowed me good cover for observing birds along the river-way and in the overhead vegetation.

Tow Path Along C&O Canal

While the Yellow-breasted Chat and Ovenbird are birds of the thickets, the equally bright **Yellow Warbler** is a bird of the high foliage. Even when it is not obvious, its spirited song is sure proof of its presence. Only about half the size of a Chat, male Yellow Warblers are totally yellow, including the underside of their tail. The exception is their finely, chestnut-streaked breast and small, coal-black eyes. They sometimes are called "canaries" or "yellow-birds."

Their song is a loud, clear, flutelike, varied whistle. It is easy to imitate, although they are rarely fooled enough to come down from their high perches to investigate. Peter Vickery, in *The Audubon Society Master Guide to Birding*, described its song as a "lovely, cheerful song; usually 3-4 well-spaced "tseet-tseet-tseet sitta-sitta-see."

The Yellow Warbler is a neotropical migrant that only nests in the U.S. and spends the greater part of its year in tropical forests where it associates with a variety of other wintering songbirds. For one example, I recall a particular Christmas Bird Count in Catemaco, Veracruz, that I wrote about in *Birder's Mexico,* as follows:

> I was surprised at the great variety of birds encountered inside the forest canopy. This isolated patch of rain forest apparently had attracted many species of birds and other wildlife that had been displaced when the adjacent forest had been cut, and I was the beneficiary of their short-term residency. I was most surprised to find an abundance of North American warblers. Magnolia Warblers were most numerous, but Orange-crowned, Nashville, and Wilson's were also abundant. In addition I recorded a few Black-and-white, Yellow-rumped, and Hooded warblers; American Redstarts and Yellow-breasted Chats; and lone Worm-eating, Yellow, and Kentucky Warblers, and an Ovenbird; I had found a Lucy's Warbler and Northern Waterthrush earlier along the stream.

Wilson's Warblers were one of the more common wintering warblers found on those Christmas Counts. I recorded it in every one of those bird parties. It is a perky little bird with a bright yellow underside, olive-colored back, a noticeable black cap, and a yellow face with black eyes. Its summers are spent in the northern portion of the United States, through all of Canada, and almost all of Alaska. It nests in dense, moist woodlands, bogs, willow thickets, and streamside tangles.

Lassen Volcanic N.P. by Betty Wauer

I also found Wilson's Warblers nesting at Lassen Volcanic National Park in northern California. One early May morning I walked the circle trail from the campground, alert for whatever birds might appear. The most abundant and most obvious was the perky Steller's Jay. It had been the first bird seen on entering the park, and since then there had hardly been a time when one or several were either visible or heard. Their loud, raspy "shack, shaak, shaak" calls and a dozen other jay sounds seemed to permeate the environment.

At Lassen's Hat Lake, the surrounding vegetation contained several riparian-loving birds: Calliope Hummingbird, Downy Woodpecker, Willow Flycatcher, Tree Swallow, Warbling Vireo, Wilson's and MacGillivray's Warblers, and White-crowned and Song Sparrows. Most numerous were Wilson's Warblers, singing rapid songs which Wayne Peterson, in *The Audubon Society Master Guide to Birding*, calls a "staccato chatter dropping in pitch at end: 'chi chi chi chi cheet chet.'" Males were especially active that morning, singing from various posts about their territories, and responding immediately to any trespasser or atypical sound. Two males came within a few feet when I spished; their all-yellow underparts and solid black caps were obvious.

Rocky Mountain N.P.

I also found several Wilson's Warblers in Endovalley in Rocky Mountain National Park. I wrote about one morning in my chapter on that marvelous national park in *The Visitor's Guide to the Birds of the Rocky Mountains National Parks*, thusly:

> The willow habitats along the valley floor were filled with bird song: Red-winged Blackbirds sang the loudest, their "konj-la-ree" songs echoed across the flats; Lincoln's Sparrows were most numerous; Wilson's Warblers were also plentiful; Black-capped Chickadees called their "chick-a-dee-dee-dee" song with regularity; several Song Sparrows were heard as well; and smaller numbers of Common Snipes, Willow Flycatchers, Black-headed Grosbeaks, Brewer's Blackbirds, and Fox Sparrows were also detected.

The **Yellow-throated Warbler** sings a rather unique song in that the syllables become faster as they descend and end with an abrupt higher note, written as "tee-ew, tew-ew, tew-ew, tew-ew, tew-wi." My most memorable observation of this lovely warbler occurred at Abrams Creek in Great Smoky Mountains National Park. I left my vehicle in the parking lot and walked along the waterway to where I found a variety of songbirds. But I was most attracted to the musical songs of a Yellow-throated Warbler singing among the sycamores.

It took me several minutes to find that exquisite bird; it was singing from the upper foliage and was mostly hidden among the large leaves. Once I located the bird, I was able to see its bright yellow throat and chest bordered with coal black sides that extended onto the cheeks and below the eyes. It had a white eye line and dark gray back with a greenish-yellow patch, like that of the Northern Parula. It also had a snow-white belly and two white wing bars.

I also found a Yellow-throated Warbler at Chickamauga Battlefield in Tennessee. I wrote about that occasion in *The Visitor's Guide to the Birds of the Eastern National Parks*, thusly:

> In summer, another warbler will nest in the park, the Yellow-throated Warbler. It, too, will nest in the pines. But west of the Cumberland Plateau, it prefers sycamore trees for nesting. This species uses a creeping motion when feeding, often in the upper foliage...It has the longest bill of any of the warblers. And its song is very distinct; a loud, ringing sound like that of an Indigo Bunting or Waterthrush.

The yellow-throated Warbler, like so many other songbirds, is a neotropical migrant. It nests in the southeastern United States, from east Texas to the East Coast, and it winters in South Florida and south to Costa Rica and east into the Greater Antilles. There it frequents palm trees, often along a particularly beautiful white sand beach. I have often thought it has a choice existence, summering in the northern forests and wintering along the Caribbean beaches.

Indiana Dunes National Lakeshore

Indiana Dunes is the extreme northern edge of the Yellow-throated Warbler's breeding range. I once found it there, feeding in the tall sycamores. Because it has an affinity for sycamores, it once was known as "Sycamore Warbler." And I later learned that this warbler has the longest bill of any of the wood warblers, enabling it to feed under bark and in deep crevices, from streamside and cypress swamps to pine and oak forests. It is a true southern warbler, summering only in the Southeast and north to the Chesapeake Bay area. In winter it prefers the southern climes from Florida and south Texas into Mexico and Central America.

Orange-crowned Warblers are nondescript yellowish birds that can be reasonably common nesters within various habitats in the U.S. I have recorded nesting Orange-crowns at numerous locations, including aspen groves in the high mountains. At Zion National Park, for example, they are a "fairly common summer resident and migrant." In *Birds of Zion National Park and Vicinity*, I wrote the following about Orange-crowned Warblers:

> Summer residents are limited to the aspen groves and brush-covered slopes in the high country. Dennis Carter and Bruce Moorhead observed an adult feeding an immature bird in Potato Hollow on June 29, 1963, and I found a territorial pair in the aspens near Blue Springs Lake on June 1,

1965. Post-nesting birds and/or fall migrants may appear in the lowlands by mid-August, and there are scattered reports to the end of October.

One week in June I hiked the Bear Canyon Trail to the highlands of the Guadalupe Mountains in Texas. I camped overnight in The Bowl, that contains a relict but declining forest. I was surprised to find the number of breeding birds there. These included Common Nighthawks; Whip-poor-wills; Mountain Chickadees; Pygmy and White-breasted Nuthatches; House Wrens; Hermit Thrushes; Warbling Vireos; Orange-crowned, Yellow-rumped, and Grace's warblers; Western Tanagers; and Dark-eyed (Gray-headed) Juncos. Breeding Orange-crowned Warblers sang a high-pitched staccato trill; their call note is a sharp *chip*. In winter, Orange-crowns join flocks of other songbirds in southern forests.

Orange-crowned Warbler

One winter, while camping at Bentsen-Rio Grande State Park in the Lower Rio Grande Valley, I experienced a truly unexpected encounter between two Orange-crowned Warblers. Mark Elwonger and I had just returned from a birding trip into Mexico; we had driven my VW van and were staying in the park's campground directly across the roadway from a birdfeeder. The weather had turned extremely cold on our trip and we were forced to return to the States earlier than planned.

That evening at Bentsen, after we had acquired a campsite, we were sitting in the van looking out the window watching what birds came to the feeder. An Orange-crowned Warbler suddenly flew in and began nibbling on a piece of orange that had been provided by the park staff. Suddenly a second Orange-crowned Warbler flew onto the feeder and it too began to nibble on the orange. However, the initial Orange-crown would not allow the second bird to feed, and viciously attacked it. It flew at the imposter and began pecking it about its head. We watched as the imposter attempted to fight back, but it was soon losing the fight. It appeared to us that it had been seriously wounded; it actually fell off the feeding tray onto ground. The first bird continued its attack even when the imposter tried to fly into the adjacent shrubbery. By this time we were watching the fight with our binoculars, watching as the first bird reigned blow after blow on the imposter. We watched as the first bird actually pecked an eye out of the imposter. Finally, after about ten minutes, the fight ended when the imposter could not long defend itself any longer. The first Orange-crown flew back to the feeder to feed. By dark, it was obvious to us that the imposter was dead.

The **Black-throated Gray Warbler** is one of my all-time favorite warblers. I became very familiar with this warbler while working in Santa Fe and living in nearby Bandelier National Monument; my wife worked there as the park's administrative officer.

Black-throated Grays are a lovely black-and-white bird except for the tiny yellow spot in front of each eye. And its song is very different from that of other warblers. Its song has been described as a simple, pleasing song of four or more notes; the last syllable may ascend or descend: 'swee.swee, ker-swee, sick' or 'wee-zy, wee-zy, wee-zy, wee-zy, weet."

While living at Bandelier, I initiated a study on the impacts of feral burros on the park's breeding birds in May and June 1977. I surveyed two comparable mesas within the park: Frijolitos Mesa possessed burros and the adjacent Frijoles Mesa (control site) was free of burros. After selecting comparable locations, marking each with flagging that divided my one-mile-long transects into 52 intervals (an area comprising 100 acres), I surveyed the birdlife in each area in the mornings on eight occasions. I recorded 46 breeding bird species on Frijoles Mesa and 32 on Frijolitos Mesa.

Bandelier N.M., Frijoles Canyon

My eventual report stated that "the greater environmental deterioration of the Frijolitos Mesa habitat was clearly a direct result of the influence of the feral burros." And when assessing the impact of burros on individual bird species, I found that although ground-nesting species incurred the greatest impact, numbers of tree-nesters, such as Black-throated Gray Warblers, also suffered from a deteriorated habitat. Reduction of various herbs and low shrubs, certainly resulted in reduced insect populations.

I also recorded Black-throated Grays at Mesa Verde National Park. That park's landscape is dominated by a pinyon-juniper woodland, a favorite habitat for Black-throated Grays. I wrote about that area in *The Visitor's Guide to the Birds of the Rocky Mountains National Parks*, thusly:

> My favorite pinyon-juniper bird is the little black-and-white Black-throated Gray Warbler. Males possess an all-black head, tiny yellow spot in front of each eye. Females are duller birds, without the coal black throats. These are active birds that are usually detected first by their very distinct, but rather quiet, songs: a buzzy "weezy, weezy, weezy, weezy-weet." They

> seem to sing most adamantly during the warm dry afternoons. Black-throated Grays are typical Neotropical migrants that nest in the United States and spend their winters in warmer climates from southern California and Arizona south to southwestern Mexico.

Alexander Sprunt, Jr., in Griscom and Sprunts's *The Warblers of America*, described their song as "A simple, pleasing song of four or more notes, the last syllable may ascend or descend; 'swee, swee, ker-swee, sick' or 'wee-zy, wee-zy, wee-zy, wee-zy, weet.'"

Another interesting fact about Black-throated Gray Warblers is that they are less "nervous" than most warblers. They go about their business in a methodical and deliberate manner, very much like vireos. I once was able to watch a pair foraging from only a few yards distance. They seemed to ignore me totally. I was reminded of an instance in Griscom and Sprunt's warbler book: W. L. Kinley, on discovering a nest, wrote:

> The moment the mother returned and found me at the nest she was scared out of her senses. She fell from the top of the tree in a fluttering fit. She caught quivering on the limbs a foot from my hand. But unable to hold on, she slipped through the branches and clutched my shoe. I never saw an exaggerated case of the chills. I stooped to see what ailed her. She wavered like an autumn leaf to the ground. I leaped done, but she had limped under a bush and suddenly got well. Of course I knew she was tricking me! But I never saw higher skill in a feathered artist.

Zion N.P., Great White Throne

While working at Zion National Park, the **Grace's Warbler** became my highlight warbler, not only because of its colorful appearance, but also because it seemed to prefer the park's less disturbed places. I usually located this warbler in the taller pines by its rather subtle songs. It is not loud, but it is a series of melodic chips. Roger Tory Peterson, in his *A Field Guide to Western Birds*, described the songs as "Cheedle cheedle che che che che, etc. (ends in trill)."

It also was common at nearby Bryce Canyon; ponderosa pines dominate the upper plateau as well as in a few scattered niches within the canyon. I recall once hiking down among the hoodoos, and the only bird sounds, besides the occasional Canyon Wren, were Grace's Warblers singing in the high foliage.

The Grace's Warbler, although it usually is necessary to see it through binoculars, spends most of life high overhead in pine trees. I recall watching a Grace's Warbler foraging for insects; it seemed to creep along the branches. And every once in a while it flew out to capture a flying insect. It truly is a splendid warbler!

The Grace's Warbler is a small warbler with a brilliant yellow throat and breast, a bold yellow streak over each eye, gray back, white belly with black streaks on each side, and two white wing bars. Although it spends its breeding season in the conifer forests of the southwestern mountains, it spends its winter months in the forests of southwestern Mexico and Guatemala. I once found one in Guatemala where it is was associated with several other neotropical songbirds; the flock was feeding among the high foliage of various pines and firs.

Over the years, I have spent an enormous amount of time with **Colima Warblers**. It is a Mexican species found nowhere else but Big Bend National Park in the United States. Since I already have written a considerable amount about this very special bird, I simply will include my earlier discussion below.

Colima Warbler

The song came from the oaks over the trail ahead. It was a rapid series of melodic chips that ascended slightly and ended in a sharp tic note. It came again, lasting not more than two or perhaps three seconds. Then silence, except for the other nearby bird songs. I walked slowly along the trail to where I was much closer to where I had heard the song, to where I could see movement among the oak leaves just ahead. And then, almost on cue, it sang again. And there, among the deep green oak leaves, was a gray and yellow Colima Warbler, the bird of my search. As I watched, it put its head back and sang again, a spirited song that, to me, represented the essence of Big Bend National Park and the Chisos Mountains.

That initial experience with the Colima Warbler occurred many years ago, even before I was assigned to Big Bend National Park as Chief Park Naturalist in 1966. It was my first sighting of one of the most wanted of North America's songbirds, an avian specialty that birders travel from all parts of the country, even many parts of the world to see.

Until 1928, the Colima Warbler was a Mexican species only, known only from a handful of specimens. The first of these came from the state of Colima in 1889 – hence its name – utilized in the initial description by Outram Bangs in 1925. Frederick M. Gaige of the University of Michigan expanded the bird's range into the United States in 1928, when he collected one in the Chisos Mountains. Since then it has been recorded there regularly, although it has never been found to nest elsewhere in the United States. The Colima Warbler and Big Bend's Chisos Mountains have become synonymous.

The first scientific name used by Bangs in his initial description was *Vermivora crissalis*, but the genus name was later changed to *Halminthophila*, and it was changed again to *Leiothylgis*, and a third time to *Oreothlypis*. However, most bird books still use *Vermivora*. At least nine North American warblers fall into the *Vermivora* genus. These include Blue-winged, Golden-winged, Tennessee, Orange-crowned, Bachman's, Nashville, Virginia's, and Lucy's. The later three species are most similar to the Colima in general appearance and song. In fact, some ornithologists claim that Colima and Virginia's interbreed. The Virginia's Warbler migrates through the Big Bend region and nests in the Rocky Mountains as far south as the Davis Mountains in West Texas, only about 150 miles north of the Chisos Mountains. It differs from the Colima by possessing a yellowish instead of a gray chest, and a smaller bill.

The Colima Warbler is a relatively large warbler (4.5-5 inches in length from bill to the end of the tail) with a heavy bill (for a warbler), snow-white eye rings, brownish-gray plumage with a yellowish rump and under tail coverts, and a reddish crown patch. Its spring plumage, especially a territorial male singing in the sunshine from an oak in Boot Canyon, is truly a lovely creature. From below, its gray chest that contrasts with its dull yellow flanks and bright yellow crissum is unique.

Sooner or later, all active birders will visit the Chisos Mountains to record the Colima Warbler. But it is not an easy bird to find. Besides the extensive travel often necessary even to reach West Texas, the Chisos Mountains, the centerpiece of Big Bend National Park, is more than 100 miles south of the nearest major highway (US 90). And then, once one enters the national park and drives up into the Chisos Basin, one must hike at least three miles into the Pinnacles or to Boot Canyon, a nine-mile round-trip hike, to find a Colima Warbler. The Pinnacles Trail is the nearest but steepest route to find Colimas; that trail climbs about 1,800 feet in three miles.

Boot Canyon, Big Bend N.P.

Boot Canyon is located in the heart of the Chisos Mountains. This moist highland drainage lies along the southern edge of Emory Peak, the park's high point at 7,835 feet elevation, and runs for about two miles from the South Rim (7,200 ft.) to the Boot Canyon "pouroff" into upper Juniper Canyon. Boot Canyon is dominated by Arizona cypress and Arizona pine, with an understory of Emory and Grave's oaks and mountain maples. Mexican pinions, alligator and drooping junipers, gray oaks, and Texas madrones dominate the adjacent slopes.

During most years, Boot Canyon and its side-canyons support up to 35 pairs of Colima Warblers in spring and summer. Several other areas in the Chisos are utilized by smaller numbers of breeding Colimas. For instance, the oak-maple canyons along the north sides of Emory Peak, such as below Laguna Meadow and along the Pinnacles Trail, support nesting birds most years. Birds also usually can be found in upper Pine Canyon, on the eastern side of the mountains, and accessible by an eight-mile backcountry road and a two-mile hike. And at least since 1995, a dozen of more pairs of Colimas have utilized a high north-facing slope in upper Green Gulch.

Birds of Chisos Woodlands

Associated birds of interest to visiting birders within the Colima Warbler's preferred environment include Blue-throated and Broad-tailed Hummingbirds, Acorn and Ladder-backed Woodpeckers; Cordilleran and Ash-throated Flycatchers; Hutton's Vireo; Mexican Jay; Violet-green Swallow; Bushtit; Bewick's, Canyon, and Rock wrens; Painted Redstart (most years); Black-headed Grosbeak; and Hepatic Tanager. The painting above, by Nancy McGowan, contains the principal birds found in the upper Chisos Mountains.

All the *Vermivora* warblers (except for Lucy's) are ground-nesters. Colimas construct their nests of grasses and hair along a gradual slope, usually well hidden among the grasses or under a root or rocky outcropping. Adjacent woody vegetation provides natural stair-steps for arriving and leaving the nests, as well as adding protection from predators or accidental damage from passing hikers. Occasionally a nest is built so close to the trail that a birder can peak into the nest without ever leaving the trail; see below.

Colima Warbler, nest site

Colimas arrive on their Chisos Mountains breeding grounds during March and April. The earliest sighting is March 18. But much depends upon the availability of caterpillars and various other insects among the early leafing oaks. Tiny green oak caterpillars are especially important as a diet for the nestlings. Colimas forage primarily in the upper foliage but also seek insects in the undergrowth.

If adequate food is available, males immediately select a territory and begin to defend those areas from other Colimas. Although their singing is most vigorous during the morning and evening hours, a territorial male will often sing throughout the day as it feeds and courts its mate among the oaks and maples. Both sexes share duties during the nest-building stage as well as during incubation and feeding of the young. Early fledglings may be out and about by mid-May, although late nesters, due to a late season or early nest destruction, may not produce young until July or even August. Early departures can occur as early as August, but Colimas have been found as late as September 19. They winter in southeastern Mexico, among the oak woodlands that are not too different ecologically than those in the Chisos Mountains.

Big Bend's Chisos Mountains are the northern tip of the Colima's breeding range that extends southward into the Mexican states of Coahuila, San Louis Potosi, and southwestern Tamaulipas. Essential habitats within various mountain ranges are similar, all at mid-elevations and dominated by oaks and maples with adjacent pines, junipers, and madrones. Shrubs, grasses, and succulents (agave and cacti) are present in the understory. The Colima's Mexican range is even less accessible than that in the Chisos Mountains, making the West Texas birds the best bet for most birders.

Chisos Mountains, Big Bend N.P.

Because Colimas breed nowhere else in the United States, and since the population there is subject to various catastrophes, such as wildfires, deep drought, and various other threats, it was considered for listing as endangered in 1967. IUCN, the International Union for Conservation of Nature, already listed the Colima as "near threatened," because of its "small U.S. population" and found only in the extremely northern edge of its range. I found that listing totally unnecessary and fruitless. To ward off that action, I initiated a survey of breeding Colimas to prove that they were common in the Chisos and also to provide an adequate baseline to allow for long-term monitoring of the populations.

That first year, I asked help from several birding friends – Jon Barlow, James Dick, Ned Fritz, John Galley, Wes Hetrick, Ted Jones, Jim Lane, Anne LeSassier, Dick Nelson, Mike Parmeter, Kent Rylander, and Francis Williams - and I set the count date to correspond to what I had found to be the peak of the Colimas' breeding cycle, the second week of May. I selected twelve areas in the Chisos Mountains to be surveyed. The twelve areas included Upper, Middle and Lower Boot Canyon, East Rim Canyon, the South Rim Trail, Mt. Emory Trail, Laguna Meadow Canyon, the Southeast Corner of the Chisos Basin, Pine and Maple Canyons, and Kibby Spring. Talleys for the first (1967) count produced 46 birds. In 1968, 65 individuals were counted, 83 in 1969, and 59 in 1970. I then skipped three years, but undertook counts within the same locations in 1974 and 1976. In 1974, 48 were counted and 52 were found in 1976.

I later analyzed all my data and found that the Colima Warbler population in the Chisos seemed to be in a direct correlation to the amount of precipitation during the breeding season of the previous year. And I was able to demonstrate that the Colima population in the Chisos was stable and viable. Being in a national park, where full protection from habitat degradation was constant, listing as endangered or even threatened was unnecessary.

Like Colima Warblers, known to nest in only a small area of the U.S., the range of **Golden-cheeked Warblers** also is limited. It nests only within a reasonably small portion of the Edwards Plateau in central Texas. This 35-county region, from west of San Antonio to the Dallas-Fort Worth area, is rectangular shaped. In *Birding Texas*, co-authored with Mark Elwonger, we included the following paragraph about the Edwards Plateau:

> The Edwards Plateau, often called the Hill Country, comprises about 37,500 square miles in central Texas. The Balcones Escarpment forms a distinct boundary on the southeast and south, while the northern and northeastern borders blend gradually into the Northern Plains and Brush Country regions. Natural vegetation of the Edwards Plateau includes climax grasses on the open slopes and valley bottoms; the rocky slopes and breaks area support live and shinnery oaks, junipers, and mesquite. Brush and juniper species have invaded much of the region's grasslands and open savannahs.

The Golden-cheeked Warbler is the "official state bird of Texas." Because of its unique status in Texas, it is subject of annual censuses by a cadre of volunteers. Although there are no official population numbers, data reveals that there has been a gradual decline. Human populations within the range of Golden-cheeks have increased by 50% between 1990 and 2010. Land uses have doubled; Golden-cheek numbers have declined an equal amount.

Golden-cheeked Warbler, by Greg Lasley

I have had numerous observations of Golden-cheeked Warblers throughout their range. In *Birding Texas*, we mentioned several locations where Golden-cheeks can be found. These include Balcones Canyonlands National Wildlife Refuge, Barton Creek Habitat Preserve, Dinosaur State Park, Emma Long Metropolitan Park, Fossil Rim Wildlife Center, Friedrich Wilderness Park, Garner State Park, Guadalupe River State Park, Longhorn Caverns State Park, Lost Maples State Park, Pedernales Falls State Park, Possum Kingdom State Park, and Meridian State Park. My favorite sites is Meridian State Park in Bosque County. I included a photograph of "Betty Wauer listening to Golden-cheeked Warblers along Shinnery Ridge, Meridian State Park" in *Birding Texas*.

Golden-cheeks are habitat specific in that they are found only in areas of mature Ashe junipers within various juniper-oak communities. Nesting birds utilize strips of juniper in their nest construction, lined with rootlets, feathers, and hair, secured by spider webbing, and camouflaged by bark strips.

> Paul Ehrlich and colleagues included a paragraph on Golden-cheeks, thusly: Breeding: Stands of mature Ashe juniper ("cedar brakes"). 1 brood. Displays: Courtship inconspicuous; male fluffs feathers, gives "chip" notes, occ faces female and spreads wings. Female chooses site. Built in 4 days. Diet: Entirely insectivorous.

Golden-cheeked Warbler, by Greg Lasley

During the almost four years working in Great Smoky Mountain National Park, I was able to hike the Appalachian Trail on numerous occasions. The trail provided me with a superb perspective of the Smokies and its birdlife. One of my favorite Smoky Mountain birds was the **Black-throated Green Warbler**. I wrote about one encounter in my Great Smoky Mountain chapter in *The Visitor's Guide to the Birds of the Eastern National Parks*, thusly:

Not far ahead was Double Springs Gap; beech, yellow birch, yellow buckeye, and a few other trees dominated the low point on the ridge. I stopped to admire the yellow flowers of a bluehead lily, one of the park's many unique plants. Then, almost overhead, the husky notes of a Black-throated Green Warbler attracted my attention. An adult male sat among the lower branches of a birch tree, singing a song that is one of the hallmarks of the highland forests of the Smoky Mountains.

The song of the Black-throated Green Warbler is very distinct – a slow, drowsy series of five to eight notes, "zee zee zee zo zee," ascending at the end. Although my bird moved off into the woods, its song continued all the while I was exploring the beech gap habitat. Ornithologist Arthur Bent pointed out that the Black-throated Green is one of our most persistent singers and that one individual "gave 466 songs in a single hour and more than 14,000 in 94 hours of observation." And Frank Chapman wrote that its song has "a quality about it like the droning of bees; it seems to voice the restfulness of a summer day."

Black-throated Green Warbler, by Greg Lasley

The Black-throated Green is one of the most attractive warblers of the Great Smoky Mountains. A breeding male sports a coal black throat and breast, a yellow face and an olive line through the eye, olive green upperparts, two white wing bars, and a white belly. The female is a duller version of the male.

I had another rather unexpected encounter with a Black-throated Green Warbler at Prince Albert National Park in Saskatchewan, Canada. I included that incident in my Prince Albert National Park chapter in *The Visitor's Guide to the Birds of the Central National Parks*:

> I was attracted to the rather swinging, but short, "wee-o wee-o wee-che" song of a Magnolia Warbler. I found this little bird gleaning insects from spruce boughs. Its bright yellow underparts were heavily streaked with black, which formed a larger spot on its bib, and I also could see its tail distinctly marked with a broad white band. Then, higher up to the right was another, darker warbler. I spished slightly to entice it into the open, and suddenly I was enveloped in warblers. Four Black-throated Greens, a Blackburnian, a Black-and-white, and a Magnolia were all in front of me, chipping excitedly over my presence. I remained perfectly still, and within 5 to 10 seconds they resumed feeding in a reasonably open view.

Fundy National Park, located in New Brunswick along the Atlantic Coast, provided me with a greater number of fall warblers than anywhere else during our Canadian adventures. The park's headquarters area seemed to be the center of the fall migrants. I wrote about that area as follows:

> Hundreds of Pine Siskins, those mites of the coniferous forests, were already busy as I walked through the campground. Their thin, ascending trills were everywhere, blending with the wispy songs of the Golden-crowned Kinglets and the toy horns of Red-breasted Nuthatches in the adjacent forest. Suddenly, a couple of dozen warblers swept across the trees and landed in the open birches at the entrance station: Tennessee, Black-throated Blues, Black-and-whites, and others that I missed when they abruptly moved on toward the coast.

Bay-breasted Blackpoll Warblers

Warblers made up the bulk of my sightings at Fundy. Black-throated Green and Tennessee Warblers were the most numerous, but I also found substantial numbers of Yellow, Magnolia, Yellow-rumped, and Black-and-white warblers. And smaller numbers of Nashville, Northern Parula, Chestnut-sided, Cape May, Black-throated Blue, Blackpoll, and Bay-breasted warblers. American Redstarts and Common Yellowthroats were also seen. In addition, Yellow-bellied, Alder, and Least flycatchers, Solitary Vireo, and Northern Oriole were present in what seemed to be one huge bird-party. I watched dozens of these southbound birds as they went about finding sufficient food to fuel their six- to eight-week journey to tropical climes where they would spend the winter.

Pine Warblers are one of the few warblers that reside full-time in the U.S., rather migrating to the south each winter. It spends most of its time in the treetops year-round. But it is not an easy bird to see well, even with binoculars; I have had a sore neck on many occasions. Finding one often requires one to listen for its song, which is a clear, musical trill, a slow succession of soft notes with little variation, which it sings continuously. Its song has been compared with the faster and higher-pitched Chipping Sparrow song but softer.

Pine Warbler

The Pine Warbler is well named because it rarely ventures out of the pines. An adult male has yellow underparts, olive green upperparts, yellow eyebrows, two prominent white wing bars, and rather obvious white spots at the corners of the tail. Females have a similar pattern but are duller. While feeding, it has a habit of creeping about the pine needles, often near the very tops of the trees. And Pine Warblers also join wintertime bird parties, but they usually stay in the pines even when the rest of the party is feeding in the lower foliage.

One of my earliest encounters with Pine Warblers occurred during a visit to the Texas Big Thicket, then a semi-undisturbed area near Houston. While working at Big Bend National Park, I had been asked to evaluate the area for possible inclusion within the National Park System. I spent several days walking and boating the area. I eventually recommended the Big Thicket be added.

During my surveys, I kept a record of all the birds observed, and one of the most numerous species recorded was the Pine Warbler. It seemed to me that almost every one of the tall shortleaf pines provided a viable habitat for resident Pine Warblers. A few of its neighbors included the Acadian Flycatcher, American Redstart, Brown-headed and White-breasted Nuthatches, Carolina Chickadee, Tufted Titmouse, Red-eyed Vireo, Northern Parula, Northern Cardinal, and Pileated Woodpecker.

And finding **Swainson's Warblers** in the lowland thickets was most exciting. I recoded this warbler along the Turkey Creek Trail, just south of the Gore Store Road. That Swainson's Warbler was a lifer that day!

> Mike Austin described the Swainson's Warbler's song as a "loud, ringing song that can be heard from mid-April through May. The songster uses a variety of perches, from ground level to about eight feet up. It prefers the densest cover and, because it usually cohabits with the cottonmouth and a variety of thorned bushes, commando-type excursions to view it is not recommended."

Its song, described by Paul Sykes, in *The Audubon Society Master Guide to Birding*, is a loud and clear melody, "tee-o, tee-o, whit-sut-say, bee-o, tee, toot-sut-say, bee-o." And Frank Chapman added that its song "has an indescribable, tender quality that thrills the senses after the sound has ceased."

Big Thicket Sign, by Betty Wauer

Nuthatches are small passerine birds that are characterized by large heads, short tails, and powerful bills and feet. Most species exhibit gray to bluish upperparts and a black eye stripe. They are non-migratory and live in their habitat year-round. They forage for insects hidden in or under bark by climbing tree trunks and branches.

I was also impressed with the tiny **Brown-headed Nuthatches**. They seemed to be everywhere! I later wrote about these fascinating birds in my "Big Thicket National Preserve" chapter in *The Visitor's Guide to the Birds of the Central National Parks*:

> Brown-headed Nuthatches were flying from pine to pine, walking up and down the scaly trunks, and gleaning the long pine needles. All the time they were uttering thin "pee-you" calls in constant communications with their neighbors. This nuthatch is one of the few birds recognized for its use of tools; it uses a flake of bark like a crowbar to pry up other pieces of bark to retrieve insects underneath. Unmated males are known to assist paired birds with feeding nestlings and fledglings.

I also found these little birds at Gulf Islands National Seashore in Mississippi. One of my most productive birding sites at Gulf Island was around the park's headquarters. The picnic area at the end of the parking lot was best. The tall longleaf pines apparently were favored by Brown-headed Nuthatches. These little tree huggers walked up and down the trunks, sometimes upside down on branches, searching for insects. On one of my wintertime visits there, I also recorded Red-bellied and Downy Woodpeckers; Yellow-bellied Sapsucker; Northern Flicker; Eastern Phoebe; Blue Jay; Fish Crow; American Robin; House Wren; Tufted Titmouse; Carolina Chickadee; Ruby-crowned Kinglet; Pine Warbler; Northern Cardinal; Spotted Towhee; Chipping, Song, Swamp, and White-throated sparrows; and American Goldfinch.

Brown-headed Nuthatches, as well as Red-cockaded Woodpeckers, also occur in Big Cypress National Park; they utilize very similar habitats as those within the Big Thicket. I found both species along the Florida Trail near the Oasis Ranger Station. Past logging of the pine stands had affected both species, although both are still present in reduced numbers.

North America claims 22 species of woodpeckers. While they share many characteristics, each species can be quite unique! They range from small to large and plain to colorful. They are known for their powerful beaks, long tongues, sometimes flashy colors, and their excellent climbing ability. Most feed on insects, but four feed on sap. Some live in forests or woodlands, while others live in deserts. Woodpeckers are a versatile family of birds, and some are personal favorites!

Of the 22 species of woodpeckers, only two are found exclusively in the United States. The **White-headed Woodpecker**, found only in the western states of Washington, Oregon, and California, is distinctly marked with its all-white face and chest, and black back and wings with white primaries, most visible in flight.

And the **Red-bellied Woodpecker** is found throughout the eastern half of the U.S., and, like the smaller Ladder-back, has black-and-white barring on its back, pale underparts (slight reddish belly), and gray cheeks and red cap and nape.

The Downy, American Three-toed, and Black-backed Woodpeckers share their range with Canada.

Red-cockaded Woodpeckers, like Brown-headed Nuthatches, also occur in pine stands in the Texas Big Thicket. It is listed as "threatened" or "endangered" due to declining numbers wherever it resides. Therefore, all during the time I spent in the Big Thicket, I watched for "candle trees," evidence of its presence.

Red-cockaded Woodpeckers occur in open stands of mature longleaf pines (80-120 years old) whose heartwood has been destroyed by a fungus. They excavate cavities into the living tree; a colony may use the same hole for several decades. Resin "wells" are excavated and maintained around the cavity entrance so that the resultant sticky resin repels predators such as rat snakes. Because the dripping resin resembles wax, the trees are called "candle trees." Although a colony can include up to 30 cavity trees, only one cavity is used by a single pair of breeding birds. Other mature birds serve as helpers.

Author on trail at Big Thicket, by Betty Wauer

Gary Clark included a description of Red-cockaded Woodpeckers in his *Book of Texas Birds*, thusly:

> The Red-cockaded Woodpecker is surely the most fascinating of all woodpeckers found in Texas. Only slightly larger than the Downy Woodpecker in your backyard, the Red-cockaded Woodpecker has a black-and-white body with horizontal barring across it back, with white cheeks, and a black cap and nape. The male has a nearly imperceptible streak of red feathers of the side of its head just behind the eye that makes it appear to be wearing a little red cockade.

Red-cockaded Woodpecker populations have declined drastically throughout their range, even after being listed as endangered in 1970. The reasons for their decline vary from habitat removal to overprotection of habitat, such as fire suppression. When the surrounding understory gets tall enough to allow predators to reach the nest, these woodpeckers dessert the site. Paul Ehrlich and colleagues, in *Birds in Jeopardy*, point out that the current range is only 10 percent of the original and the world's population is "about 7,400 individuals."

The National Park Service has developed a fire management program for the Big Thicket that is designed to restore and maintain certain areas, such as the Hickory Creek-Savanna Unit, for the woodpecker. The birds are now extremely rare, and protection of the few remaining colonies is of utmost important.

I also recorded Red-cockaded Woodpeckers in Congaree Swamp National Monument in South Carolina. I added additional information acquired from that visit, thusly:

> Red-cockaded Woodpeckers maintain rather large territories and spend a good deal of time away from their roosting sites. They feed almost exclusively on wood-boring insects and beetle larvae in pine trees, often spiraling upward on the trunk and to the crown. However, they also feed on the ground, among berry thicket, and even on standing corn, from which they extract corn earworms. Terres reported that a family of Red-cockaded Woodpeckers "may eat up to 8,000 earworms per acre" during summer.

The **Acorn Woodpecker** is my favorite woodpecker! I first got acquainted with this noisy bird during my teens while living in western California. Its West Coast range extends from the Oregon-Washington line south to Northern Baja. It also occurs from Arizona and New Mexico southward within Mexico's Sierra Madre Occidental. And it also occurs in the Sierra Madre Oriental, from the northern edge in the Texas Big Bend Country, all the way to northern Columbia, South America. It is widespread in oak woodlands throughout.

I know it best from the Southwest, especially Big Bend National Park. In *A Field Guide to the Birds of the Big Bend*, I wrote that:

> It is a common resident of the Chisos woodlands; abundant at Boot Springs; the Acorn Woodpecker is most numerous in the upper canyons and ranges into the lower edge of the oaks in late summer and fall...There are several nesting records, all in live oaks or piñons from early May through July. The easiest place to find this bird is within the canyons along the north side of the Basin Campground.

Pinyon-Juniper Woodland, Big Bend N.P.

At Boot Spring, a tall snag was a favorite resting place for these woodpeckers that flew out and back or were chasing flying insects. Several dozen acorns had been wedged into holes the previous year. Acorn

Woodpeckers store acorns each fall, jamming them into holes drilled for that purpose and retrieving them to eat during the remainder of the year. There are records of old trees with fifty thousand or more storage holes, but none of the various trees I saw that morning contained more than a few dozen acorns.

I also wrote about an experience with Acorn Woodpeckers *in For All Seasons, A Big Bend Journal*, thusly:

> May 2 (1994). Rain had fallen last night throughout the Chisos Mountains. The Pine Canyon pouroff had changed overnight from the dry cliff face that I had observed on April 24 to a lovely waterfall. Dozens of butterflies and birds were taking advantage of the available water to drink, shower, and, apparently, simply congregate. During the two hours that I sat below the falls this morning, I observed five species of butterflies and sixteen bird species visiting the waterfall.
>
> Acorn Woodpeckers were the most numerous of the resident birds. They seemed to be having a wonderful time. I counted eight individuals flying back and forth from adjacent trees, and several times they bathed in the fresh, flowing water while clinging to the cliff. Mexican Jays were present as well, but these birds seemed disturbed by my presence and stayed at the water only for only brief periods before flying away with sharp scolding calls. A pair of Black-headed Grosbeaks flew in together, remaining for about ten minutes, and then departed together. A lone Common Raven flew by, circled the upper cliff, and then flew off with loud cawing; apparently my presence kept it from coming closer. All the while, a Band-tailed Pigeon sat at the clifftop, apparently waiting for me to leave.

Acorn Woodpecker

I had another encounter with Acorn Woodpeckers one fall morning at Chiricahua National Monument that I will long remember. It began with a troop of Mexican Jays that were feeding on oaks on the numerous oak trees at the Faraway Ranch. They were everywhere, actively searching the two to three dozen oak trees, and raising a clamor by their loud calls and the loud drumming of falling acorns on the tin roofs. Standing beneath the trees, I watched one individual collect an acorn and hammer the prize with its large, heavy bill until it was able to retrieve the rich meat.

I was so engaged watching the jays collect and open acorns that I did not notice the two or three Acorn Woodpeckers that were also gathering acorns. But instead of immediately opening the acorns, they collected two or three and flew over to a nearby pine tree where they shoved the acorns into pre-dug holes. They placed them there for long-term storage to when they needed a fresh supply.

Acorn Woodpecker, storage holes, Betty Wauer

On another morning at Chiricahua, I hiked to Echo Canyon. Birdsong from the full-time residents as well as the summer-only residents was all around me. Mexican Jays were heard now and again, but their general lack of dominance that day suggested that they still were involved with nesting chores. Acorn Woodpeckers, however, adequately filled in for the usually noisy jays. These woodpeckers were not only vocally active but also easily observed, flying here and there, and paying little attention to a hiker. I wrote the following morning in *Birding the Southwestern National Parks*, thusly:

> Their loud calls, usually described as "jacob" or "wack-up," never ceased. I discovered a pair of these gregarious birds on a tall snag just off the trail, and I was able to examine them at leisure. The Acorn Woodpecker is a middle-sized woodpecker with an all-black back, tail, and wings, except that in flight their white wing patches and a snow-white rump are obvious. Its most conspicuous feature is its contrasting black, white, red, and yellow head, almost clownlike in appearance.

Paul Ehrlich and colleagues provided some additional information on Acorn Woodpeckers in *The Birder's Handbook*:

> Live in communal groups up to 16, consisting of at least 2 breeding pairs plus their young of previous nestlings and cousins. Large clutches result of 2 females. Reproduction highly dependent on size of acorn crop. In CA, maintain all-year communal territories, with communal acorn stores. In AZ, some nest as lone pairs and migrate if insufficient food is stored; some AZ populations do not hoard. Young independent at ca. 22 months. Oft evicted from nest cavity by starlings. Attack squirrels, jays, nuthatches, titmice, esp Lewis' Woodpecker (which also store acorns) that raid caches.

The range of the **Golden-fronted Woodpecker** covers all of Texas and much of Mexico and extending south to Nicaragua. Most of my encounters have been in Big Bend National Park and/or the Lower Rio Grande Valley; it is common in both localities. In South Texas it can be found in both its natural habitat and in suburbia. Its abundance there no doubt is due to its broad diet to take advantage of a wide diversity of foods.

I am most familiar with it from Big Bend where it is a "rare spring migrant; rare nesting bird along the river," according to *A Field Guide to the Birds of the Big Bend*. I wrote the following:

> The status of this bird has changed recently. Before 1970, it was considered a rare spring migrant, and there were two fall sightings...From late 1981 through 1983, however, there were twice as many sightings as there had been previously...Apparently, the species has discovered the available habitats within the park. The nearest previously known nesting sites were at Calamity Creek, just south of Alpine, and at The Post, a park just south of Marathon.

Since those early years, Golden-fronted Woodpeckers have become abundant in the park's lowlands, especially at Rio Grande Village and Cottonwood Campground, both along the Rio Grande. Cottonwood trees dominate both sites.

I have found it foraging high on trees as well as on the ground. Although they principally take insects and other arthropods, they also utilize whatever fruits may be in season. Tim Brush, in *Nesting Birds of a Tropical Frontier* wrote:

> Golden-fronted Woodpeckers may be among the most important frugivores for particular plant species. I have seen some unusual foraging methods and sites in the Valley; May 26, 1997: One repeatedly captured moths near boat ramps at Bentsen, from tree trunks and branches. June17, 1997: 3 foraged on mud and moist algal mat at Willow [the pond at Santa Ana].

Golden-fronted Woodpecker, by Greg Lasley

Rylander, in *The Behavior of Texas Birds*, added a note on the Golden-fronted Woodpecker's feeding behavior: "Omnivorous, feeding on insects, nuts, berries, acorns, and a wide variety of other food items. They probe for insects beneath the bark of tree trunks, less often foraging on the ground or capturing moths in flight. They cache food in bark crevices."

Although Golden-fronts can usually be found in the proper habitats with little searching, their loud, rather constant calls, help to locate it. I recall one occasion at Santa Ana National Wildlife Management Area when I was attracted to a bird among a brushy area rather than in the open as normal. It apparently had found some food near the ground, and it seemed to be either shouting in joy or calling its neighbors. That call was a surprisingly soft, rolling, but scolding *ker r wood r r, r*, repeated several times.

Brush also wrote about the bird's range expansion: "During the twentieth century, Golden-fronted Woodpeckers expanded their range about 2500 miles, into the Texas Panhandle and southwestern Oklahoma, and west into the Big Bend region of Texas. Their numbers may have increased in coastal areas after trees were planted."

In some ways, the Red-headed Woodpecker is the icon woodpecker. Its image is so often used on plates, placards, and such, many people think first of a Red-headed Woodpecker when discussing woodpeckers. That is quite understandable because it is such a marvelous bird. That is despite its limited range; breeding Red-heads occur only in the eastern half of the United States and Canada. Those of us living in the West never are blessed with this truly outstanding bird. In addition, Red-heads are migratory, residing in the Great Plains and eastward to the Maritime Provinces in summer. They overwinter in the Southeast, from east Texas, through most of the southeastern states to the East Coast and Florida.

Red-headed Woodpeckers rarely need to be described. It is the only woodpecker with an all-red head, snow-white chest, belly, and rump, and coal-black wings. In flight they show their black wings and tail with bright, white inner wing patches. It truly is a remarkable bird!

My most memorable encounter with the Red-headed Woodpecker occurred while camping at Cottonwood Campground in Theodore Roosevelt National Park in North Dakota. I wrote about that occasion in *The Visitor's Guide to the Birds of the Central National Parks*, thusly:

> That morning, the most outstanding bird at Cottonwood Campground was the Red-headed Woodpecker. Red-heads are one of the most distinct of all the woodpeckers; males possess snow white underparts, rump, and inner wing patches, coal black wings and tail, and a brick red hood...I counted at least six males during my walk around the campground loops; the females were undoubtedly sitting on nests out of sight. But the males were flying from tree to tree, calling hoarse "kweeer" or "kwee-arr" notes. Arthur Bent, in *Life Histories of North American*

Woodpeckers, included a quote about this bird by John James Audubon: "With the exception of the mocking-bird, I know of no species so gay and frolicsome. Indeed, their whole life is one of pleasure."

I also found Red-headed Woodpeckers at Mammoth Cave National Park in Kentucky. They were significant members of park's avian community, and were some of the park's most obvious and easiest bird to find and watch. Seven individuals were found within the picnic area right next to the busy visitor center parking lot. They were obvious and active one early morning in July. Their loud calls vibrated throughout the area, and I found three adults calling from the very tops of snags that stood above the high foliage. Five youngsters were present as well, acting like any other youngsters enjoying the early morning. And I enjoyed their unusual behavior; they seemed to fall from the foliage out of control, then suddenly glide to another snag or patch of foliage. The immature birds lacked the bright red heads of the adults, but their behavior, even as immature birds, already copied the adults.

Northern Flickers were present there as well, and once I observed a young Flicker join the Red-heads. At least three adult Flickers called out occasionally; their nasal "kee-yer" calls seemed almost to express enjoy of the youngster's frolics with the Red-heads.

The smaller **Ladder-backed Woodpecker** is another woodpecker that I became acquainted with at Big Bend. It primarily is a desert woodpecker, although it occasionally is found in the lower edge of the pinyon-juniper woodlands. It seems to be a very adaptable bird; I have found it nesting in cottonwoods, willows, oaks, pinyons, century plant stalks, utility poles, and wooden signposts. Rio Grande Village and Cottonwood Campground are sure bets for finding Ladder-backs. And during Christmas Bird Counts, totals of 22, 16, 32, 24, and 34 were found on the Rio Grande Village Counts from 1966 through 1970, respectively.

Ladder-backed Woodpecker at nest

Ladder-back's range is extensive, from central Texas to southeastern California and southward into Central America. True to their name, this seven-inch woodpecker sports black-and-white barring on its back and tail, spotted sides, and a black-and-white face pattern, not unlike that of the Bridled Titmouse. Males possess a bright red crown. Ladder-backs excavate cavities in various trees, shrubs, and even tree cactuses, such as saguaros and cardons. I added the following in *Birding the Southwestern National Parks*:

> Nature effectively provided a three-species construction crew to build apartment complexes for distantly different groups of wildlife. The builders use the cavity for only one year, pecking out a new one annually. The construction, however, is normally done in summer or fall, after the nesting season. For the saguaros, the postseason construction of new cavities allows the cactus to form a callus over the soft tissue inside the cavity by the following season. Saguaro cavity nesters, therefore, rarely utilize nest lining.

Two additional woodpeckers live within the saguaro community and take advantage of the cavities: Gila Woodpecker and Gilded Flicker. The **Gila Woodpecker** is a middle-sized woodpecker with its total range corresponding with that of saguaros and the larger cardons, a similar tree cactus south of the border. During one of my many earlier visits to Saguaro National Park, I had discovered that it is next to impossible to walk among the saguaros without seeing and hearing this bird. A Gila Woodpecker is easily identified by the black-and-white barring on its back, rump, and central tail feathers; gray-tan head and underparts, except for the yellowish wash on the belly; and the male's red cap. In flight the bird shows small but bright red wing patches. It calls with loud "chuur" notes.

Saguaro Cactus

The **Gilded Flicker** is equally synonymous with saguaros, the common tree cactus of the Sonora Desert. Its range runs from the southern tip of Nevada, much of southwestern Arizona, southward into Mexico's Baja Peninsula, and east of the Sea of Cortez in Sonora. The Gilded Flicker is a somewhat smaller yellow-shafted flicker-look-alike with a larger black chest patch. It differs from Northern Flickers by its whitish rather than gray cheeks.

Saguaro cavities, which are used for both nesting and roosting, have the additional feature of being cool in summer and warm in winter, varying from outside temperatures by as much as ten to fifteen degrees. They also have the advantage of holding relative humidity that is five to ten percent higher than in the outside air. This significantly lessens the drain on the bird's body moisture and is a particular advantage for nestlings.

Almost twenty species of birds are known to utilize saguaro cavities for nesting. Besides woodpeckers, they include the American Kestrel; Western Screech-owl, Ferruginous Pygmy-owl, and Elf Owl; Ash-throated and Brown-crested Flycatchers; Western Kingbird, Purple Martin, Violet-green Swallow, Cactus, and Bewick's Wrens, Bendire's Thrasher, Western Bluebird, European Starling, Lucy's Warbler, House Finch, and House Sparrow.

I also found Ladder-backs at Joshua Tree National Park in California, located in the Mojave Desert. Although there are no saguaros there, nesting Ladder-backs utilize cavities in Joshua trees and a few other small trees and shrub. In fact, Ladder-backs are the park's most characteristic bird. A male on the Barker Dam Trail sat in the open for an unusually long time; Betty was able to take several minutes of videotape as we admired its black-and-white features. This is the smallest of two typical woodpeckers that live in the Joshua tree forest, taking insects from infected trees and excavating nest holes for themselves and a variety of other cavity nesters.

The larger, less numerous **Northern Flicker** excavates larger cavities where it and several other species of wildlife nest in season. Flickers spend considerably more of its feeding time on the ground, where it hops and walks about, very un-woodpecker-like. Flickers eat more ants than any of the other North American woodpeckers, utilizing insect larvae, acorns, nuts, and grain. Both Ladder-backs and Flickers are important members of the Joshua tree community.

Joshua Tree

Pileated Woodpeckers are significantly larger than Flickers and more than twice the size of Red-cockaded and Ladder-backed Woodpeckers Although Pileated Woodpeckers do not occur in desert habitats, they do frequent the same pine habitats as Red-cockaded Woodpeckers. But, instead of residing in only small areas of the southeastern U.S., like Red-cockaded, their range extends over all the entire eastern half of the country, in a broad band across the northern states and southern Canadian provinces, as well as along the West Coast of North America.

While growing up on the West Coast, in Petaluma and Santa Rosa, California, I became very familiar with this bird, and when working along the Russian River, Pileated Woodpeckers were reasonably common within the forested areas from the lowlands up onto the higher slopes.

California Forest

Much later, after my 32-year career working for the National Park Service, Betty and I visited every national park in the U.S. and Canada; we encountered Pileated Woodpeckers on many, many occasions; I wrote about some of those encounters. For instance, at Big South Fork National River and Recreation Area in Kentucky and Tennessee, I wrote the following in *The Visitor's Guide to the Birds of the Eastern National Parks*:

> The pileated woodpecker is one of the world's most outstanding birds. Not only is it the largest of our woodpeckers, looking very much like Woody Woodpecker of cartoon fame, but it also is one of the most vocal. And Bandy Creek Campground, at Big South Fork, has one of the largest populations of these wonderful birds of any place I know. Their loud, raucous calls can be heard there at any time of day.

A pair of Pileated Woodpeckers (according to Webster's *Ninth New Collegiate Dictionary*, both "pie-le" and "pill-e" pronunciations are correct) were perched on a huge black oak tree very near our campsite. Their large, black bodies, bright red crests, and the bold white line that extends from their white cheeks down their necks was obvious. The male, evident from the red patch behind the bill, suddenly called out a loud, high-pitched, and rather nasal "kuk-kuk, kuk-kuk." It then took flight, looking almost like a huge bat. Its great, broad wings beat slowly but powerfully enough to drive this large bird in a direct line (although they often exhibit an undulating flight), giving the distinct impression that it was composed mostly of wings. Its flight is so different from that of most other birds that it usually can be identified at a great distance. It is similar to American Crows in both flight and size.

Pileated Woodpeckers require big trees and large acreage. They have disappeared from many parts of their range due to timber cutting and fragmentation of their habitat. They utilize year-round territories and may mate for life. Nest cavities are constructed in dead trees from 15 to 85 feet above the ground but usually 28 to 35 feet high. Their diet consists primary of insects that they dig from dead and dying trees, usually leaving huge, gaping holes from their efforts. Their winter diet may also include fruits and nuts, especially during colder winters when most insects are no longer available.

The female Pileated Woodpecker at Bandy Creek Campground remained in place near our campsite, working the deep hole in the oak tree. It probably had found larvae of beetles or some other wood-boring insects, and I watched as it continued to feed. I could only imagine its specialized tongue, equipped with tiny barbs to aid in capturing insects, probing the cavity it had dug with sharp pounding of its heavy bill.

The Big Thicket was once the homeland of the **Ivory-billed Woodpecker**, but it is now considered "extinct." The last known specimens were taken in Liberty County in 1904, although almost every year since there have been rumors of sightings. In fact, several years ago, I spent several days following up on rumors that were circulating about recent sightings; I obviously was not successful.

Oberholser and Kincaid described Ivory-bills as follows:

> Raven-sized black woodpecker with a huge ivory-colored bill; yellow eyes; pointed crest (mostly scarlet in male, entirely black in female). A white stripe begins on each cheek and extends down side of neck until the two stripes more or less come together on mid-back. In flight, both

surfaces of flight feathers (except outer most primaries) white...Usual calls: high-pitched nasal yaps – *kent, kent, kent, kent.*

There are, however, two equally large woodpeckers in Mexico – **Lineated Woodpecker** and **Pale-billed Woodpecker**; I have found and enjoyed both on several occasions. They look pretty much alike, black bodies with a red head and a white streak running from the neck onto their back. The white stripe continues below the eye to the yellowish bill on the Lineated Woodpecker, but it does not extend beyond the neck on the Pale-bill, and its bill is heavier and ivory colored. In addition, Lineated shows a distinct red crest with a black patch behind each eye; Pale-bill's head is all-red except for a black eye-patch.

I learned early-on how to identify both species audibly, just by hearing them drum on a tree trunk. Lineated's drum is a hard and rapid, 2-3 seconds in duration; Pale-billed's drum is a distinctive, loud, rapid double-rap, less often up to seven distinct double-raps in a rapid, resonant series. Both species occur within about the same range in Mexico, along both the Atlantic and Gulf Coasts and all of the Yucatan Peninsula.

Imperial Woodpecker

I must also mention my quest for the apparently extinct **Imperial Woodpecker**. I discussed my search for this large bird in Mexico's Maderas del Carmen in *Birding Mexico*. I include the entire sequence of events as follows:

> I spent almost the entire day wandering on the forested plateau. The timber was tall once again, but I could not help imagining what it might have been before the loggers came and changed things. I searched that day for trees that had been used by a woodpecker larger than those I already had seen throughout the forest. Acorn Woodpeckers were common within the lower woodlands and ponderosa pine stands, and an occasional Hairy Woodpecker was found, too. Northern Flickers were present in the highlands, as well. But I was looking for evidence that the Imperial Woodpecker, North America's largest, existed within the Maderas del Carmen.
>
> The initial idea of these mountains containing Imperials began on my first visit to the Maderas del Carmen in 1969. The Mexican highlands had enticed me ever since 1966, when I first went to Big Bend National Park as chief park naturalist. My hopes of seeing the del Carmen highlands finally transpired when six of us climbed the steep burro trail above Los Cohos Mine into a different world. We camped at Los Cohos Spring and hiked out each day to explore the scenic wonders and learn what we could about the local flora and fauna.
>
> One afternoon, high above camp on the western rim, I found three ponderosa pine snags with large, oblong holes in the trunks twenty-five feet above the ground. I was surprised at the size of these holes and took several photographs. I assumed at the time that the holes were remains of Pileated Woodpecker activities. Since that bird had never been recorded in these mountains, I intended to document their presence. It was not until later that I learned that Pileated Woodpeckers do not occur in Mexico. The only large woodpecker that frequents habitat like that of the Maderas del Carmen was the endangered Imperial, a bird known only from the Sierra Madera Occidental, far to the west. Like the Ivory-billed Woodpecker of the southeastern U.S. lowlands, the Mexican Imperial is more an enigma than a reality. And yet, there in the Maderas del Carmen was possible evidence of its existence far out of its previously known range.

Maderas del Carmen, Forested Highlands

Three weeks later I returned to the Maderas del Carmen highlands to try to find the bird or hard evidence of its existence. I found additional nesting trees, although every one was old and not adequate. I attempted to climb to one of the nest holes, hoping that an ancient feather had been left behind, but a near fall reduced my enthusiasm for that method of discovery. If I had known then that those trees were the last bit of evidence I was to find, I would have risked my neck again and again until I retrieved any clues that remained in the nesting cavities.

One more piece of circumstantial evidence came my way on the second trip. I met a bear hunter, wondering along Madera Canyon early one morning. We struck up a conversation and he soon was telling me about the local wildlife. I learned about the bears and *panteras*, as well as the deer and fox. I opened my Peterson field guide to the plate on hawks and falcons and asked him if any one of those birds lived in these mountains. With only a moment's hesitation he pointed to the Cooper's hawk, Goshawk, and Peregrine Falcon. Although I was a little uncertain of the Goshawk then, I later found it nesting in Madera Canyon. I next turned to the plate on western warblers. Again, he was correct. He pointed only at the Painted Redstart and Colima and Olive Warblers.

The only picture I had found of the Imperial Woodpecker at that time was in Ernest Edward's 1968 bird-finding book. I had photocopied that plate and it had reproduced quite well. I took that copy from the back of my book, unfolded it, and asked the bear hunter if any of the birds on that plate lived in the Maderas del Carmen. He looked at all the illustrations. Then he pointed at only one, the Imperial Woodpecker. I asked him how recently he had seen one. He said it had been a long time, maybe four of five years ago; he told me he used to shoot them for food because they made a very good meal.

I have returned to the Maderas del Carmen seven times to search for the Imperial Woodpecker, but since 1970 have found no new evidence. My last visit to the Carmens was in 1976, and with the same amount of anticipation as I had experienced on that second trip. Two friends – Joan Fryzell and Grainger Hunt – had told me of seeing a "large crested woodpecker" in Canon del Oso that previous year. Joan, Grainger, and I revisited the site in 1976, but I again returned without proof.

Living in Texas, I can reliably find eleven woodpecker species within the state: Acorn, Downy, Golden-fronted, Hairy, Ladder-backed, Red-bellied, Red-cockaded, Red-headed woodpeckers, and Northern Flicker are resident, while Yellow-bellied and Red-naped Sapsuckers can be added during most winters. I have had the most experience with the Acorn, Golden-fronted, and Ladder-backed Woodpeckers. All three are common in the Big Bend area; Red-bellied is the woodpecker I see where I live in Brazos County, in the eastern half of the state.

North America is home to 19 species of owls which represent virtually every North American habitat. These owls range from the alpine tundra to the southwestern deserts, and from sea-level to above tree line in the highest mountains.

Owls are fascinating birds. Their intense, mysterious gaze, their often-strange behavior, the thrill we get when we realize we've spotted one sitting in a tree or taking flight are special. Here are some of my most memorable owl encounters.

As a cavity-nester, the tiny, nocturnal **Elf Owl** is seen only after dark. Considered the world's smallest owl, it weighs less than two ounces, about the size of a small sparrow. Besides their tiny size, they possess a dark bill, whitish eyebrows, and bright yellow eyes that give them a ferocious appearance. And it often is common within areas of both the Sonoran and Chihuahuan Deserts.

Despite its tiny size, it has a surprisingly loud voice and utters "chucklings and yips like a puppy dog." Although most of its calling is done at dusk from the entrance to the nest chamber, and again at dawn, they also call incessantly during moonlight nights in spring, after which they are difficult to detect. Elf Owls normally sleep during the daylight hours, and spend each night searching for prey. These can include an amazing variety of tiny creatures, including insects caught on the wing or on the ground, such as moths, grasshoppers, and crickets, plus scorpions, lizards, and small snakes. I remember watching one Elf Owl at Big Bend, right after one of my evening programs at the Chisos Basin Amphitheatre, chasing moths around the overhead entrance light before turning it off for the night.

Elf Owl, by Greg Lasley

Most Elf Owl sightings are limited to flashlight observations of a head at the entrance to a nest hole high in a saguaro or various other structures, including trees, utility poles, and even buildings where a woodpecker had drilled a compartment. It seems that woodpeckers and Elf Owls have a strange relationship; one designs the house and the other takes residency.

I also wrote about Elf Owls at Big Bend in *For All Seasons, A Big Bend Journal*, thusly:

> March 23 (1968). I heard the first Elf Owls of the year at the Graham Ranch (at the west end of Rio Grande Village) during an evening visit there. Their strange warble – eight to ten low whistles like "hew-ew, hew-ew, hew-ew, hew-ew, hew-ew, hew-ew, hew-ew," – was a welcome reminder that the new nesting season had begun. In spring and summer, these tiny owls can be surprisingly common throughout the lowlands and up into the lower edge of the pinyon-juniper woodlands. They are especially evident when they first return from their Mexican wintering grounds. Although many of the adults undoubtedly return to the same territory in consecutive years, subadults and others wander about a good deal in search of suitable nesting sites. Nest cavities can occur in a wide range of places, such as natural cavities or deserted woodpecker nests in utility poles, fence posts, and numerous trees and shrubs. Once a suitable territory is located, the male will spend most of his time calling from singing perches along the edges. But once a female is attracted and mating occurs, both birds are far less vociferous and more difficult to locate.

Over the years, I have become involved with the study of two other little owls, both more than twice the size of the Elf Owl: Ferruginous Pygmy-owl and Flammulated Owl. My encounter with **Ferruginous Pygmy-Owls** was the result of the managers of the King Ranch in south Texas becoming interested in ecotourism and asking Paul Palmer, an active birder and history professor in nearby Kingsville, to access the Pygmy-owl population on the ranch for a possible tour business. My involvement occurred when Paul, a long-time birding friend, asked me to participate in his surveys. I wrote about our surveys in my *Raptors* book, thusly:

> We began our survey on the Norias division of the King Ranch, the southwestern quarter of the ranch, on November 2, 1989, and we found one individual that very first day. We had stopped along the roadway in a habitat we assumed was the most likely to possess resident Pygmy-owls. We climbed out of the vehicle and stood there for several minutes hoping to hear a Pygmy-owl. Finally, without hearing our bird, we began imitating its rather distinct call. And almost instantly we got a response. A Ferruginous Pygmy-owl was responding about fifty to sixty feet away among the mesquites.

Although we did not see the vocalist that first day, we soon realized that to find Pygmy-owls on the Norias was simply driving the various roadways and stopping to listen for their easily recognized calls, a series of harsh and inflected "poip" notes, often 10 to 60 consecutively, and followed by a 10-second lull before being repeated. I later discovered that male call notes alternate between the "poip" notes and clear whistles. Female calls are higher pitched and more aspirate. Its vocalizations have also been described as a mellow series of "whah" notes, like the sound made by blowing across the opening of a partially-filled bottle of water.

Ferruginous Pygmy-Owl

These little owls are an overall gray-brown to ferruginous (reddish-brown) color, hence its name. A frontal view reveals white underparts with ferruginous streaks, a long tail with brown to blackish bars, yellow eyes with black pupils in a gray-brown face, with eyebrows bordered with whitish lines, and a rounded head without any evidence of ear tufts. The nape contains a pair of black, oblong eyespots that serve as an extra set of "eyes" that is said to "fool" potential predators.

Our November 2, 1989 finding of an Elf Owl on the King Ranch in a sense opened the door to the discovery of a major population within the United States. Following that initial visit, I worked out arrangement with the King Ranch management to establish survey transects all during 1990 and 1991, and I also was able to fly the area with Ansa Windham on January 3, 1991. That flight allowed me a better perspective of the entire area. And eventually I developed a rough estimate of the total Pygmy-owl population. I later wrote an article about my activities and findings, co-authored with Paul and Anse, that was published in *American Birds* in 1997.

Prior to my King Ranch Pygmy-owl encounters, I had seen this tiny owl on numerous occasions in Mexico, where it can be surprisingly common throughout the country. My first-ever sighting was along the Río Corona in Tamaulipas. Walking along the river we heard its mellow, staccato whistles repeatedly, rapidly but monotonously; we finally discovered it perched on an acacia in full view from less than fifty feet. It remained there long enough for us to see it well before it flew off into a brushy area where it disappeared.

Years earlier, while working in Zion National Park, I got acquainted with another little owl: **Flammulated Owl**. I later summarized its presence in the park in *Birds of Zion National Park and Vicinity*, thusly:

> *Otus flammeolus*. Rare migrant and a probable summer resident. This species was not known from the Zion region until one was caught in a mist net at a banding station in Oak Creek Canyon, May 8, 1964; it was banded, photographed (see photo below), and released. Another was found perched on a willow in the Watchman Residence Area the same day; it was also banded and released. Still another was found dead in Zion Canyon, May 11; the specimen was preserved. These records were all obtained during a period of unseasonably cold, stormy weather. An additional record is a bird seen on a Douglas fir along the south fork of Taylor Creek, May 21, 1964 (Wauer). A fifth Flammulated Owl was caught in a mist net at the Springdale Ponds, May 7, 1965, one year after the first record during a similar period of inclement weather; it too was banded, photographed, and released. It is likely that this owl nests at higher elevations in the area.

Flammulated Owl, banded

Until my numerous reports about Flammulated Owls during stormy periods in spring, this owl was thought to be a full-time resident in several western mountain ranges. But its migrant status was eventually acknowledged. And to add additional evidence for migration, it was even found on oil platforms in the Gulf of Mexico: Brian Gibbons (per. com.) recorded one on a platform 20 miles offshore from Port O'Connor on October 10, 1999. The following day one was found 310 miles east of Louisiana. And in 2000, Gibbons recorded another one on the Port O'Connor platform on October 15.

After almost four years in Zion, I transferred to Big Bend National Park where I was again involved with Flammulated Owls. In *A Field Guide to the Birds of the Big Bend*, I added the following:

> Fairly common summer resident (March 30-September 24) at localized highland areas. The Flammulated Owl has been found most often in Boot Canyon where I saw a juvenile with a captured monarch butterfly on June 8, 1968. During April and May it can usually be called up right after dark with a few hoots that need only partially resemble the deep *boot* call of this bird. One must stay overnight at Boot Spring, however, to be

assured of seeing it. It begins its nightly activities about one hour after sunset – approximately 10:00 to 10:15 p.m. in May and June – and is usually found in the main canyon just below the cabin. It also has been seen on the north slope of Casa Grande.

After retiring from the National Park Service, Betty and I moved to Victoria, Texas, near where her four boys were living at the time. Our property, about 15 miles outside of downtown, containing a large backyard where I built several planters for a variety of flowering plants designed to attract hummingbirds and butterflies. I also built a small pond and a watering system for a birdbath set in the ground. That birdbath became a popular drinking and/or bathing place for a variety of wildlife: Armadillo, Gray Fox, Striped Skunk, Ringtail, and several birds, including Red-shouldered Hawks and Barred Owls.

Barred Owl, bathing

I was able to photograph several of those bathers out of my kitchen window, which was only about 25 feet away from the birdbath. The tree-lined Guadalupe River was about one mile from my property, and I occasionally heard the deep, distant hooting of Barred Owls. And they seemed to thoroughly enjoy bathing in my yard.

And the local Red-shouldered Hawk was also a regular at my birdbath. I understood why Red-shoulders have often been described as the "neighborhood hawk." They, like Barred Owls, nest in the neighborhood, and their springtime calls, as they circle their territory, are something special. This fascinating buteo is well-marked with a white breast with numerous reddish bands, white banded back, a black-and-white banded tail, and reddish shoulders. Its legs are bare and yellow. In flight, it shows reddish forewings, black streaks on the trailing edges, and black wingtips.

Red-shouldered Hawks also were commonplace during my visits to the Big Thicket area of Texas. I recall one occasion when I was able to study one well-marked bird as it fed on a small rodent while it perched on an open tree limb. Its rufous shoulders, barring on its chest and breast, and banded tail were obvious. An adult male, perched in the morning light is a gorgeous creature.

Additional songbirds, usually listed as perching birds or passerines, contain the most diverse group of all birds. They have four toes, with three pointed forward and one backward, which helps them to perch on branches and other surfaces. And they all sing to communicate and/or to attract mates. Examples include flycatchers, vireos, wrens, warblers, orioles, and finches.

One of my all-time favorite songbirds is the **Phainopepla**. I have recorded this lovely creature in both the Chihuahuan and Sonoran Deserts. Sonoran Desert birds seem to utilize more arid habitats than Phainopeplas in the Chihuahuan Desert. Sonoran birds range from central California to all of the Baja Peninsula and throughout much of the Southwest. Chihuahuan Desert birds occur from the Texas Big Bend Country southward throughout most of central Mexico. Secondly, it is the only North American member of the Family Ptilogonatidae. Its closest relative is the Gray Silky-flycatcher, a Mexican endemic.

There are few songbirds with the character and appearance of the Phainopepla. It is impossible to misidentify; no other songbird possesses an all coal-black body, a tall black crest, and bright red eyes. In flight, they show a large white patch on the outer edge of the wings. Females are gray instead of black, and show a pair of whitish wing bars.

Phainopepla

I have had numerous encounters with Phainopeplas. My most memorable sightings are those from Organ Pipe Cactus National Park in Arizona and at Big Bend National Park in Texas. Nowhere are they as abundant as they are at Organ Pipe in both winter and spring. They are especially common in desert washes where mistletoe clumps are widespread on the mesquite, ironwood, and paloverde trees; this charismatic bird seemed to be present at almost every clump. It usually sits at the very top of a tree or shrub for a considerable length of time, then suddenly dashes out after a passing insect in a pursuit that may extend for sixty to eighty feet. At times its flittering is so graceful as to be reminiscent of a large butterfly. During courtship, males perform fantastic display flights, circling and zigzagging high above their perspective mates. In mid-February, I watched groups of five to eight Phainopeplas fluttering together in a sort of cooperative display. After three or four minutes they returned to their respective perches among the mistletoe.

In *Birding the Southwestern National Parks*, I added the following:

Phainopepla is Greek for "shining robe," referring to the male's shiny all-black plumage. This bird is the only North American member and the only desert dweller of the Silky-flycatcher family [Ptilogonatidae]; three other species live in the highlands of Mexico and south to western Panama. All possess a tall, shaggy crest, but the Phainopepla is distinguished in being all black with ruby-red eyes and snow-white wing patches (evident only in flight). Its abundant calls are discrete mellow notes, like a low "work" or liquid "quirt," which can be heard for a considerable distance. When disturbed they utter harsh "ca-ra-ack" cries. Their song is a short series of mellow notes that prompted Allan Phillips and colleagues, in *The Birds of Arizona*, to describe it as "sweet gargling." Organ Pipe Cactus biologist Tim Tibbitts told me that the Phainopepla's song is among the most common and varied in the Sonora Desert.

High above Mazatlán, Sinaloa, Mexico, in a humid pine forest, I awoke one morning to a bird chorus that was dominated by a dozen or more **Gray Silky-flycatchers** and several Brown-backed Solitaires. That dawn chorus lasted only until the first rays of the sun touched the highest treetops. Then the level of bird song declined by at least half. Many of the Silky-flycatchers were perched on tree-lupines that dominated the edge of our campsite. And further up the mountain, we found an estimated 120 individuals perched on the higher vegetation. Awaking that morning to the amazing dawn filled with such an abundance of dawn songs was worth a lifetime.

Gray Silky-Flycatchers

Early mornings in the tropics are like nowhere else. Bird songs are by far the dominant sounds. The dawn chorus may last for as little as twenty minutes or for more than an hour, depending upon the time of year and weather. Some birds sing their territorial songs only during this brief period of the day, although they may call or sing other less expressive songs at various other times during the remainder of the day. Most bird songs are different than their calls. But to hear the largest number of bird songs at their peak, one must be awake early to experience the dawn chorus.

There are close to twenty species of native flycatchers in North America. In most cases, flycatchers have dull colors but there are some exceptions, such as the bright red Vermilion Flycatcher or the spectacular looking Scissor-tailed Flycatcher.

Insects make up their principal food items, and they are well known for their acrobatic efforts in catching flying insects in the air. At times, they stop in mid-flight and hover, and they may then pick prey from leaves and branches. Flycatchers prefer high perches where they have the advantage of seeing a larger area to find prey.

Some flycatchers are difficult to identify because they are so similar in size, color, and markings. Sometimes only the bird's call can help separate them from each other.

Most flycatchers are rather plain little birds which possess the fascinating behavior of flipping their tail. There are a couple exceptions, however, the Great Kiskadee and Scissor-tailed Flycatcher. Although they lack the tail-flipping behavior, they nevertheless are distinguished birds. And they frequent very different habitats. The Kiskadee can be found about southern wetlands while the Scissor-tail is a bird of the open country from Texas to Oklahoma and Arkansas, truly a bird of the American Southwest.

The **Great Kiskadee** can never be compared with those smaller, plain, tail-flipping flycatchers. Kiskadees are much larger and well-marked with a russet back and tail, bright yellow chest and belly, and a head pattern with a white collar, wide russet band that runs from the bill through the eyes, and a russet cap. And if that isn't enough, its call is a slow but loud and deliberate *kis-ka-dee*, with emphasis on the *kis* and *dee*. It also gives a loud *kreath* call; its name is obviously derived from its very distinct *kis-ka-dee* call.

Most of my observations of this large flycatcher have occurred in the Lower Rio Grande Valley, particularly at Bentsen-Rio Grande Valley State Park and Santa Ana National Wildlife Refuge; both areas lie along the Rio Grande. Kiskadees are partial to ponds and other wetlands. For example, near the entrance to Santa Ana is a pond on the right where I never have missed finding a Kiskadee or two. I have watched this flycatcher also feeding on minnow-sized fish that were captured by short dives from limbs hanging low over the water. Those dives, not too unlike kingfishers, often produced a tiny fish held in its bill as they returned to their original perch. On one occasion, it caught a damselfly off a stab and returned to its perch where I watched it tear off the wings, letting them drop into the water, and consume the soft body.

Great Kiskadee

Although most of my Kiskadee sightings have come from my many visits to the Lower Grande Valley in Texas, I also have recorded it on several trips into Mexico. I recall a Kiskadee on a camping trip at Rancho Nuevo, Tamaulipas. I wrote about that dawn chorus in *Birder's Mexico*, thusly:

> The loudest of the bird songs was coming from a thicket where a Spot-breasted Wren repeated its song over and over again...Off to the right was the drawn-out *who whoo* of a Red-billed Pigeon, and I could hear another a hundred yards ahead on the left. Plain Chachalacas, at least four individuals were also calling somewhere beyond. Two Brown Jays were screeching in the trees just ahead. An Olive Sparrow sang from some shrubs a few yards to the left...a Ferruginous Pygmy-owl added its repetitive single-note whistle to the chorus. I also detected at least three Masked Tityras and a Golden-fronted Woodpecker calling from the same general area. The distinct call of a Tufted Titmouse, fairly close, was almost overlooked. Further off to the left I detected an Elegant Trogon...And about three hundred feet ahead, a Great Kiskadee joined in the dawn chorus.

I also recall a morning along the Rio Cihuatán that runs between the Mexican states of Jalisco and Colima. Once a gallery forest, the streamsides on my last visit contained little more than scattered patches of what was once a significant habitat that supported a whole array of birds. I wrote about that morning in *Birder's Mexico*, as well:

> We found a Citreoline Trogon in a small patch of trees surprisingly close to the main highway. And we found a Ferruginous Pygmy-owl perched on a small acacia, and being mobbed by several other birds. Included in this party were Social and Vermilion Flycatchers, a pair of Thick-billed Kingbirds, and a Great Kiskadee.

All those flycatchers were diving and screaming at the lone Pygmy-owl. But I was most impressed by the Kiskadee that seemed to dominate the mobbing process. It made dive after dive at the Pygmy-owl, even scrapping its back once or twice. Most of the times the Pygmy-owl ducked just in time. The Pygmy-owl finally decided that enough was enough and made a fast retreat into a nearby brushy area. The Kiskadee followed it, screaming at it in kis ka dee fashion, staying after it until the Pygmy-owl was firmly hidden within the brush.

I also can personally attest to their vigorous defensive behavior. On finding a pair of nest-building Kiskadees along the entrance road to Santa Ana National Wildlife Refuge in Texas, and while trying to photograph that pair of nest-building birds, I was attacked time and again by both birds. I remember being surprised by their loud cries on each dive that continued until I moved elsewhere; one individual even followed me for a hundred feet or more, chastising me for my intrusion.

The other flycatcher that has had my full attention on numerous occasions is the **Scissor-tailed Flycatcher**. Living in central Texas, returning Scissor-tails are one of my best signs of spring. Males are first to arrive in early May, but within another few days the females appear. They can be common, flying here and there and perched on open treetops and on highway signs. I have seen individuals perched just above vehicular traffic and even dashing out for insects amid the traffic. Their flight is strong, direct, and rapid.

Scissor-tail's dawn songs begin in the very early mornings even before there is sufficient light to see them well. Their predawn song, according to

Rylander's *The Behavior of Texas Birds*, "is delicate, eerie, and mesmerizing…it begins slowly and deliberately with a dry *chit chit chit chit chit*, then gains speed and rises in pitch to end in an agitated but musical *chicka chicka ckicka CHICK.*"

Watching a courting male is amazing as it flies high in the air where it performs up and down and zigzagging flights with its long tail trailing in the sky and giving "snap calls" all the while. During those flights, with wings out and fluttering, they display their salmon-colored armpits and underwing linings. One ornithologist stated it is "an aerial ballet of incomparable grace." And occasionally they will hover in flight.

Scissor-tailed Flycatcher, painting by John O'Neil

Scissor-tails are one of our most charismatic birds. The longer-tailed, brighter males establish their territory almost immediately upon their arrival, and begin defending that territory by chasing competitors away from preferred sites, often the same sites utilized the previous year. Some chases extend for a hundred feet or more.

Rylander also stated that "Although Scissortails are usually observed singly, in pairs, or in family units, migrants sometimes roost communally in congregations containing more than 250 birds. This tendency to flock in such extreme numbers (on roosting trees) is apparently unique among our flycatchers."

Flycatcher diets consist primarily of insects which they catch in flight and sometimes gather on the ground or glean off leaves. But on their wintering grounds in southern Mexico and Central America, they also will consume berries. In June 1979, I watched a dozen or more Scissor-tails flycatching over a large open pasture along Pipeline Road in Panama. I stood there for 15 to 20 minutes admiring their amazing acrobatics. I wondered how they never crashed into one another.

There is something truly special about Scissor-tailed Flycatchers. Each spring I welcome their arrival with great excitement. Their arrival in Texas is a sure sign that the new season has begun. They seem to add a special exuberance to my otherwise rather normal existence. Few birds have the appeal of this charismatic songbird. In Texas, it is often called the "Texas bird of paradise." It also is known as Swallow-tailed Flycatcher and Long-tailed Kingbird. Whatever its name, it is extra special. And in Oklahoma it is the official state bird.

This tiny ball of fire can hardly be compared with any of the other flycatchers. The male **Vermilion Flycatcher** has no equal! In Mexico, this little bird is known as *la brasita fuego*, or "little coal of fire." Females lack the vermilion color, but they are a lovely pinkish color, nevertheless. But both possess the same lively personality. I learned to appreciate both while working at Big Bend National Park. Rio Grande Village and Cottonwood Campground, both located along the Rio Grande, contained perfect habitats, open areas with scattered trees and shrubs, that were used for nesting and perching sites for these amazing little birds.

The voice of the Vermilion Flycatcher is a loud, energetic *pit a se* that during their nesting season can be repeated several times in a row, all the while jerking its tail. I have been able to stand within a dozen feet of its nest without it flying away. They seem fearless near the nest, but at other times will fly away on my approach. When flying off, however, they may continue to sing a series of soft, sweet, twittering notes.

Vermillion Flycatcher, male, by Greg Lasley

In *A Field Guide to the Birds of the Big Bend*, I wrote the following:

> The Vermilion Flycatcher is most numerous during the spring migration from March 7 through mid-April, when it may occur up to 4,500 feet; on March 25, 1995, Carol Edwards and I found at least 45 individuals at RGV [Rio Grande Village]. There are numerous spring reports from the desert lowlands, and it regularly visits PJ [Panther Junction] from March 7 to March 25, probably the peak of the northbound movement. Fall migrants are most numerous in September; there seems to be a hiatus during early October, with another wave of migrants during the third week of October, increasing the lowland populations. A few remain all winter at RGV and Cottonwood Campground. Christmas Bird Counts at RGV annually tally two to six birds.

There truly is something extra special about Vermilion Flycatchers. It is not just the male's gorgeous plumage, but also its unique personality. I have spent a considerable amount of time with this little Flycatcher, and I have ended each visit with sadness. Like a good and loyal friend, that nearness lifted my love for birds and all the outdoors.

Warblers are some of the smallest birds found on the North American continent. They are known for their long migrations, traveling from South America and the West Indies to the northern regions of Canada and back again. These small jittery birds hardly ever stop moving, always hopping from branch to branch, scurrying along the tree trunks and limbs. There are others who live on the ground, hidden in the undergrowth, where you may only hear their songs and not see the birds that are singing them.

One of our most unusual warblers is the **Red-faced Warbler**. Although its body, with an all-gray back and wings, with one short, whitish wing bar, and white undersides, is nothing special, its head is spectacular! Its entire face and upper throat is bright red, its cap is coal-black, and it has a small white spot on the back of its head.

Its voice is a full chip or *tchip*, and its song is "a sweet warbled series, *wi tsi-wi tsi-wi, si-wi-si-wichu*, and variation," according to Steve Howell and Sophie Webb in *A Guide to the Birds of Mexico and Northern Central America.*

Red-faced Warbler, painting

Like several other tropical songbirds, Red-faced Warblers occur only as a breeding bird in the United States, and only in the mountains of southern Arizona and the southwestern corner of New Mexico. These upland areas are the northern extension of Mexico's Sierra Madre Occidental. Their U.S. breeding sites are limited to these southern sky-islands, namely the uplands of the Santa Ritas, within Coronado National Forest, about 25 miles southeast of Tucson, Arizona. In winter, Red-faced Warblers are found only in Mexico, from Sinaloa and Durango to Veracruz and Chiapas, and south to Guatemala.

Preferred habitats for these gorgeous warblers occur in pine and aspen forests, usually between 6,600- and 9,800-feet elevation. Nests, small cups constructed of leaves, grass, and pine needles, are hidden in the ground amid debris, and sheltered under a shrub, log, or rock.

My U.S encounters with this colorful bird occurred in Madera Canyon in the Santa Ritas. I stayed a couple nights in a cabin at Santa Rita lodge and hiked up-canyon into the higher forest from there. I recall locating two individual Red-faced Warblers a couple miles above the lodge and watching them feed in the foliage of oak trees. I had read about their quirky habit of flicking their tail while feeding, and so I was prepared as I watched it flicking its tail sideways while capturing caterpillars.

Red-faced Warblers have also been recorded a few times in Texas. Its presence in the Trans-Pecos region was mentioned by Mark Lockwood and Brush Freeman in *Handbook of Texas Birds*: "Red-faced Warbler *Cardellina rubrifrons.* Very rare late summer and fall migrant to the Chisos Mountains in the Trans-Pecos. Red-faced Warbler is an accidental visitor elsewhere in the state."

I provided additional details about its appearance in the Trans-Pecos in *A Field Guide to the Birds of the Big Bend,* thusly:

> The distribution and status of this brightly marked warbler was documented by Greg Lasley, Dave Easterla, Chuck Sexton, and Dominic Bartol in a comprehensive 1982 article in the *Bulletin of the Texas Ornithological Society*; which is well worth reading. They point out that the species nests in the Sacramento and Sandia mountains of New Mexico and that the Big Bend records likely are migrants. However, its presence in the Chisos should be carefully monitored. Because the habitat in the Chisos Mountains appears like that on its known breeding grounds in New

Mexico; increasing records may suggest an eastern movement of its breeding range.

The possibility of the Red-faced Warbler being a "regular" migrant through the Big Bend area is questionable. It does not occur in Mexico's Sierra Madre Oriental, the mountain range below the Big Bend area, but it is a reasonably common breeding bird in the highlands of the Sierra Madre Occidental, directly below Arizona and New Mexico. Howell and Webb state that it is a "breeder on Pacific Slope and in adjacent interior from Son to Dgo…F and U transient and winter visitor (Aug.-Apr) on Pacific Slope from Sin, and interior from cen Mexico, to W Honduras and El Salvador."

In addition, in *Birder's Mexico*, I wrote about finding one these Red-faced birds in Oaxaca:

> Driving across the high plateau, between Oaxaca City and the Gulf Coast, I stopped to check out the birds in an area that reminded me of the high arid landscape of Las Vegas, New Mexico. Patches of grass, low-growing prickly pears and scrub oaks dominated this habitat. And sure enough, the birds there could just as well have been in southern Arizona or New Mexico; Common Raven, Eastern Bluebird, Red-faced Warbler; the Vesper, Lark and Chipping Sparrows; and Lesser Goldfinch.

Howell and Webb also include a summary of its Mexican habitat: "Arid to semiarid pine-oak, and oak woodland, in winter also humid pine-evergreen forest and semideciduous woodland. Singly or in pairs at mid- to upper levels, tail often cocked and swung about loosely like Wilson's Warbler; joins mixed-species flocks."

Although most warblers are arboreal in nature, singing, nesting, and feeding in the high foliage, the **Prothonotary Warbler** is an exception. During the six years in which I worked in Washington, D.C., I discovered that one of the ways I could get away from the bureaucratic rat-race was to spend time in the field watching birds. With the help of other birders, I gradually located several super birding sites. One of those was the towpath along the C & O Canal. The towpath was high enough above the waterway and vegetation that allowed me good cover for observing birds along the riverway and in the adjacent vegetation.

One morning, as I was enjoying the scenery and the ambiance, I was startled by a very loud and emphatic "peet, tweet, tweet, tweet" song that burst forth from the edge of the canal just ahead. I knew immediately that this was the song of a Prothonotary Warbler, but it took me several minutes to locate this golden-headed songster. I was then able to admire it at will as it searched for insects among the dense willows.

There are few birds with the appeal of a male Prothonotary Warbler. Its brilliant gold, almost orange-colored head is a bold contrast to its rather heavy black bill, large black eyes, and blue-gray wings. Then, I realized that I also was watching a second bird, a female with duller plumage, that I had not noticed before. A pair no doubt, and I assumed these birds were going to nest nearby. I switched back to the colorful male just as it put its head back and sang its clear, ringing song, all on one pitch, "peet, tweet, tweet, tweet." What a wonderful song for a beautiful bird!

Prothonotary Warbler, painting

I remember one more time when I was able to watch a Prothonotary Warbler. This occasion happened during a visit to the Congaree Swamp

National Monument in South Carolina. Betty and I had been given permission to camp in one of the park's primitive sites and I was out walking on the nearby boardwalk that passes along the edge of the cypress-dominated wetland. Suddenly, the loud and penetrating song, "peet, zweet, zweet, zweet," exploded from near the boardwalk just ahead of me. A second later, a golden ball of feathers landed on the edge of the boardwalk. It stayed only a second before flying to a nearby cypress knee, where it perched a foot above the water and sang again, "peet, zweet, zweet, zweet." It was obvious why this rather tame warbler is sometimes called the Golden swamp Warbler. What an incredible lovely bird!

Prothonotary comes from the Latin word *protontarius*, for the yellow hood or robe worn by some church officials. It seemed most appropriate as I watched this gorgeous creature. It did somehow serve as the leading official of the Congaree congregation. I was but a visitor to the chapel.

Prothonotary Warblers reside in the U.S. only part of the year, from April to mid-August. The brightly colored males and their duller mates build their nests in tree cavities, often in cypress knees or abandoned woodpecker holes, a few to 30 feet high. They fill the cavity with mosses, lichens, dry leaves, and a variety of other materials. They usually construct two nests, one for the female, in which she will lay three to eight yellowish eggs spotted with dark brown, and a second nest for the male. The second or dummy nest helps to fool the common Brown-headed Cowbird, which often lays its eggs in the nests of smaller birds.

Prothonotary Warblers spend their winters in Central America and northern South America, utilizing wetland habitats similar to their breeding grounds in the states. Most interesting, however, is their use of communal roosts in winter, very unlike their behavior on their breeding grounds But, even then they aggressively defend an established territory.

The **Black-throated Blue Warbler** is, in my opinion, the most impressive of all the warblers I have encountered. And it is the most appropriate avian representative of the Great Smoky Mountains. This is because they utilize such a wide variety of habitats in summer, from the spruce-fir forests to the northern hardwoods and flame azalea thickets. The Black-throated Blue has long been a favorite of mine, not only because of its impressive plumage and unique song, but also because of its personality They are one of the tamest of warblers and often will allow the viewer to approach close enough to truly appreciate their gorgeous plumage.

The male Black-throated Blue Warbler is one of nature's most impressive creations. In the shadowy forest it appears dark grayish-blue above with a black face and throat and white underparts. But perched on a maple tree and highlighted by a shaft of sunlight, it suddenly becomes a dazzling silver-blue bird with a coal-black face and throat. Females, however, are very plain, with an olive-green back and whitish underparts. Both males and females possess a distinct white spot at the base of the primaries.

Flame Azalea

Unlike most warblers, Black-throated Blues seem to have no established niche in the forest; one time it will be in the undergrowth and the next time in the canopy. But wherever it goes on its breeding grounds, its distinct song of two to seven syllables, like "sweee-sweee-sweee-sweee," "zur-zur-zur-zree," or "chur-chur-chur-chur," with an upward slur on the last syllable, is usually audible. Another interpretation of its short song is "I am la-zy."

This warbler also may be considered a top candidate for the Smoky Mountains' representative to several Latin American countries. It spends its winters in Central America, South America, and the Greater Antilles. I have recorded this lovely creature on several of the West Indies. And I wrote about one encounter on Montserrat in *A Birder's West Indies, An Island-by-Island Tour*:

> A pair of Purple-throated Caribs came dashing out of the forest at one point. One remained for several minutes, perched a dozen yards or so away...Movement to the right caught my eye. I was certain it was a warbler but wasn't sure of the species. A few seconds later (the result of low spishing), a gorgeous male Black-throated Blue Warbler emerged from the forest into full view. It remained long enough for me to watch it feed on insects by gleaning them from the foliage and once by flying out to capture its prey on the wing. It also preened its lovely feathers and at one point bathed in tiny water droplets clinging to the leaves. The deep blues and contrasting colors were most evident when shafts of sunlight settled on the bird like a heavenly spotlight. Phenomenal!

Louis J. Halle loved warblers; he wrote about his appreciation of warblers, thusly:

> It is a slow acquisition, since most of the species are to be seen and heard only for a few days each year, and the rarer may be seen only at intervals of several years. The appreciation of birds, indeed the appreciation of all the phenomena of spring, cannot be dissociated from the accumulations of memory... The first Yellow-throated Warbler next year will be more the meaningful to me as it brings that moment in the woods back.

I also have written about two non-warbler songbirds which also express their appreciation of the new day beyond their dawn songs: Red-eyed Vireo and Song Sparrow.

I wrote about Red-eyed Vireos at New River Gorge in West Virginia in *The Visitor's Guide to the Birds of the Eastern National Parks*, thusly:

Inside the forest, the most numerous songsters, at least in spring and summer, is the **Red-eyed Vireo**. Despite its abundance, this five-and-a-half-inch bird is difficult to find. It lives among the upper foliage of the taller broadleaf trees, where it gleans insects from the leaves and smaller branches. Its song, a series of short but deliberate robin-like whistles, with a rising inflection at the end of each phrase, can be heard throughout the day. One interpretation of its' song is "You-see-it, do you hear me? do you believe it?" It also holds the record for singing more frequently than any other North American songster. One ornithologist recorded 22,197 songs during one ten-hour summer day. It also possesses a nasal "whang" call in its repertoire.

And in *The Birder's Handbook*, Paul Ehrlich and colleagues stated that Red-eyed Vireos possess "a repertoire of 40 song types; rarely sings same song type in succession."

I also recorded Red-eyed Vireos along the Buffalo River in Arkansas. Like its behavior at New River Gorge, it spent most of its days in the upper foliage; it rarely ventures into the lower layers of vegetation. A report titled "Arkansas Birds; Their Distribution and Abundance," by Douglas James and Joseph Neal, stated that in summer Red-eyed Vireos were "the most numerous birds inhabiting the oak-hickory forest in the upper part of the Buffalo National River."

Buffalo River

I also wrote about Red-eyed Vireos after one of my visits to the Buffalo River in *The Visitor's Guide to the Birds of the Central National Parks*, thusly:

> Stop anywhere along the Ozark entrance road in spring and summer, and you can usually hear three or four birds singing close by and others in the distance. Since only male vireos sing, sometimes right on the nest, one can appreciate not only their abundance, but also how this species has divided the forest. Territories are established like a giant checkerboard. Summer population studies in southern Arkansas showed a population of seventy-eight singing males per one hundred acres of forest stream bottomland, and thirty-five per one hundred acres of upland forest.

On another mid-September visit to Buffalo River, I canoed from Buffalo Point to Rush Landing, a beautiful stretch of the river with high bluffs and deep green forests. Bird migration was well underway, and many of the birds recorded were only passing through, following the riverway on their route south to their winter homes. Despite the late season, Red-eyed Vireos, maybe only late nesters, still sang to me along the river.

Song Sparrows usually frequent brushy sites at ground level, particularly along streams and at wetlands. If one is not seen immediately, it usually can be enticed into the open with squeaking or a sound produced by sucking the back of the hand. This little sparrow is rather drab, with brownish plumage marked by dark streaks that come together on the chest like a stickpin. What this bird does not have in color it makes up for in personality. It is one of the perkiest and most active of birds.

Its song has been described as a "merry chant – which has won for it its name. It is a voluble and uninterrupted but short refrain, and is, perhaps, the sweetest of bird songs. Although it possesses more than the ordinary number of songs, each begins with clear "sweet sweet sweet" notes, followed by a trill that drops in pitch and is highly variable.

Voyageurs N.P., Minnesota

I have encountered Song Sparrows on many occasions over the years. At Voyageurs National Park in Minnesota, for instance, while walking the Oberholzer Nature Trail one early morning, and watching ducks and pelicans on Black Bay, I was attracted to the nearby brushy area by the abundant songs of Song Sparrows. I later wrote in my Voyageurs chapter in *The Visitor's Guide to the Birds of the Eastern National Parks,* that "Voyageur's Song Sparrows possess light brown plumage with darker streaks; the heavy streaks on their whitish breasts form a black stickpin. When flying from one singing post to another, they pump their rather long tails up and down, as if to get cranked up for their next series of songs."

On another occasion, at Indiana Dunes National Seashore, I slowly wandered along a trail that skirted along Cowles Bog, named for Professor Henry Cowles at the University of Chicago, who did pioneering studies in plant ecology. Cowles Bog was designated as a National Natural Landmark in 1963. The trail passes along the northern edge of the bog, loops over forested dunes, and circles interdunal ponds and marshes. It provided a pleasant walk and an outstanding birding route.

Songs Sparrows were common at the start of the trail, singing songs that I was very familiar with. These little sparrows responded immediately to my spishing and came charging out to defend their territories. At one time at least a dozen Song Sparrows appeared to defend their breeding grounds. I was enchanted by their "sweet sweet sweet" songs. It was an occasion that comes to mind every time I have since thought about Cowles Bog at Indiana Dunes.

Song Sparrows were the subject of a 1943 study of birdsongs by ornithologist Margaret Nice, who discovered that an Ohio Song Sparrow sang a total of 2,305 songs during a single day in May. Its reputation for being a remarkable songster is well deserved. In fact, the bird's scientific name is *melodia*, Greek for melody or melodious song.

Although warblers are famous for their bright plumage and flashy behavior, there are several other songbirds that possess colorful plumage, a unique song, and are just as appealing to many birders and other nature lovers. Buntings are members of that special congregation.

The **Lazuli Bunting** is a bird of riverside thickets. Males are dressed in a plumage dominated by bright turquoise on its head, neck, nape, rump, and tail, cinnamon across its breast, and white upper wing bars. Their name is very fitting; it is derived from lapis lazuli, an opaque, azure-blue to deep-blue gemstone of lazurite. No other songbird contains such a lovely combination of colors. They are bright jewels among the greenery on their homeland landscape.

Lazuli Bunting

I watched one of these gorgeous songbirds one morning at Dinosaur National Monument in Utah as it searched for insects within the shrubbery. I made a few sharp chip sounds, and almost immediately a female appeared out of the saltbush thicket and chipped back at me, as if agitated by my presence.

A moment later the male Lazuli Bunting burst forth with song, as to help solidify its territory against this human intruder. I marveled at its bright and rapid song, a series of varied phrases that sounded to my ear as "see-see, sweert, sweert, sweert, see, see, swert, seer, see-see." During the 15 to 20 minutes that I watched these lovely and rather aggressive birds, the male sang from several levels: on the low saltbushes, from the lower branches of an adjacent cottonwood tree, from the top of a utility pole, and from the very top of the tall cottonwood. The female stayed among the dense shrubbery but proclaimed her territory with loud and hard chips.

I wrote about a similar experience with a Lazuli Bunting at Colorado National Monument in Colorado:

> While birding at a riparian area near the park's entrance, I recorded Gambel's Quail, Black-chinned Hummingbirds, Western Kingbirds, Black-billed Magpies, Bullock's Orioles, House Finches, Lesser Goldfinches, and Lazuli Buntings.
>
> The most colorful of these birds was the Lazuli Bunting, a sparrow-sized bird that is easily identified (at least the male is). No other bird possesses a bright turquoise (lazuli-colored) head, bluish back, white belly, and chestnut sides. The female is brownish, darker on the back and with lighter underparts, and a bluish rump. Both sexes are curious and usually can be attracted close-up by spishing. They will chip loudly and can be rather aggressive when nesting.

Although Lazuli Buntings can claim their rightful elegance with their bright turquoise plumage, they cannot be compared with the multicolored male **Painted Bunting** that possess a bright, deep blue head, greenish back, rose-colored breast and rump, and a black tail and wings. An outstanding combination of colors! It is almost gaudy.

Painted Bunting, male

A neotropical migrant, Painted Buntings arrive in Texas in early spring. It immediately established its territory and nests, and its young usually are fledged and on their own by mid-June. The female, a plain, greenish-yellow bird, does most of the household chores, while the colorful male is defending their territory. Most of this involves singing from various posts; it will vigorously chase away other Painted Bunting males when necessary. Its song is surprisingly loud (for such a small bird) and clear, a musical warble, like "pew-eata, pew-eata, j-eaty-you-too."

Gary Clark, in *Book of Texas Birds*, wrote that he calls "the painted bunting the "rainbow bird" because of its spangled colors of red, blue, and green. The male is arguably among the most beautiful birds in North America…The bird also signals to me the beauty of spring and early summer when the landscape is lambent with rainbow hues."

At Big Bend National Park, the Painted Bunting is an "Abundant summer resident at Rio Grande Village, common elsewhere along the floodplain and less numerous at springs and water areas up to 3,500 feet; rare spring migrant," according to *A Field Guide to Birds of the Big Bend.*

Males seldom remain all summer, but leave before the young are fledged and head south to their wintering grounds in Mexico and south to Panama. While in Texas, I found these colorful birds visiting my seed feeders in Victoria throughout their stay. I had two or three males and six to eight females at my feeders daily. And the males sang their sweet songs while perched in the tree just above the feeders. In *Naturally...South Texas*, I added following:

> There is a very good reason why we often see more females than males at the feeders. Painted Buntings, like many other birds, are polygamous; a single male will mate with several females. Although humans tend to question this practice, it is a very practical one for many birds, particularly for neotropical migrants that must select a territory, court, mate, and raise a family in a relatively short period of time. The females simply choose the male that is best able to claim and hold the most superior territory. A resource-rich territory will most likely provide a better chance of producing offspring than inferior one.

At Amistad National Recreation Area, located on the Rio Grande near Del Rio, Texas, Betty and I spent several hours at Lowry Spring, an isolated spring located along a paved but poorly maintained roadway northeast of Rough Canyon. We had the place all to ourselves; apparently it is remote enough to receive few visitors. Several tall Vasey oaks in the rocky drainage apparently enhanced the habitat enough to attract White-eyed Vireos, Black-crested Titmice, Blue-gray Gnatcatchers, and Summer Tanagers.

And Painted Buntings were abundant along the drainage, singing their spirited songs from various shrubs. I recorded its song that day as "clear musical couplets, like 'pew-eate, pew-eate, j-eata, you-too.'" Males perched in full view so that their contrasting, almost gaudy plumage was obvious. Through binoculars, I could also see their red eye rings and short, conical bill. A truly gorgeous creature!

While Painted Buntings are birds of the West, **Indigo Buntings** are eastern buntings that spends their breeding season at woodland clearing and borders in mountainous areas such as the Appalachians. During the three years that I worked in Great Smoky Mountains National Park, I found it to be reasonably common at the almost all the overlooks. If I did not see it at first, a very brief squeaking noise almost immediately attracted one of these gorgeous birds into the open. Usually, however, calling one into the open was unnecessary; they had a habit of singing at the edges. Their songs were loud and rollicking, like "sweet-sweet, where-where, here-here, see-it, see-it." And they will often sing the same song over and over, even in late summer when most of the other birds are silent. I wrote the following about these brightly colored birds in *Songbirds of The West, Personal Encounters*:

> The male Indigo Bunting has little competition for being the brightest and most visible bird of the old fields. This chunky, indigo-blue bird often sits in the open singing a loud and spirited song. One ornithologist stated their song is "a wiry, high-pitched, strident series of couplets, each pair at a different pitch, with the second pair of notes especially harsh: 'swee-swee zrett-zrett swee-swee zay-zay seeit -seeit."

Indigo Bunting, by Betty Wauer

I discovered that Indigo Buntings remained on their breeding grounds long after most of the other songbirds have begun their southern migration. The reason for their late departure is most likely because they often produce two broods. Three to six eggs are laid in a well-woven cup placed in the crotch of a woody shrub five to fifteen feet high. Nest-building materials include almost anything found locally: dry grasses, dead leaves, strips of bark, weeds, feathers, hair, Spanish moss, snake skins, and an assortment of human debris, from pieces of paper to shreds of clothing. The second brood may not be fledged until early fall.

My most favorite bunting is the **Varied Bunting**. It truly is a lovely bird! Although at first sighting it looks blackish with some varied colors, but in the right light it shows black lores and chin that contrast with a violet-blue face and collar, reddish nape and back, and black wings and tail, edged with bluish color. The throat and chest are dark reddish that becomes deep purple on the belly and undertail coverts. Its head and body are a veiled buffy color. Because of its marvelous features, I utilized one of Greg Lasley's super photographs of a male Varied Bunting for the cover of my book, *Feathers and Scales*, Writings About Birds and Butterflies.

The voice of the Varied Bunting was described by Harry Oberholser and Edward Kincaid in The Bird Life of Texas as "a thin, crisp, energic warbling similar to that of the Painted Bunting, but more obviously phrased and less rambling. The male, from atop a bush of yucca stalk, sometimes a high wire, sings from mid-April to July. Their call is a thin chirp chirp."

Varied Bunting by Greg Lasley

The breeding range of this lovely creature is limited to the southern portions of the southern states of New Mexico and Texas. All my observations occurred during the six years in which I worked at Big Bend

National Park in South Texas. I included an extensive discussion about this species in *A Field Guide to Birds of The Big Bend*, thusly:

> Varied Bunting. *Passerina verisicolor*. Fairly common summer resident; migrant; casual in winter. The earliest spring reports include lone birds at Hot Spring on March 25,1994 (Dorthy Hagewood), and at Rio Grande Village on April 4, 1972 (Wauer and Russ and Marian Wilson). Reports are more numerous after mid-April, including one caught in a mist net at PJ (Panther Junction) on April 18, 1969 (Wauer); it was banded and released. During the first two weeks of May, Varied Buntings became fairly common in weedy arroyos and along the roadways. Males arrive on their breeding grounds at mid-elevation first, and the females join them a few to 10 days later. On April 29, 1970, I found four singing males along the Window Trail, but no females. A male was actively defending a territory in this same area on May 24, 1968, and a male was feeding two youngsters there on July 19, 1969. I found a nest containing three nestlings in Blue Creek Canyon on June 4, 1968, and seven singing males along the one-mile stretch of Cottonwood Creek, just behind the Old Ranch, on July 13, 1968. I also found four singing males and one nest on a squaw bush along one mile of Blue Creek Canyon, just above the Homer Wilson house, on June 4, 1968. It also can be found during the nesting season in the brushy drainage below Government Springs, in the shallow drainage above Dugout Well, and on June 5, 1970, I found four singing birds among the rather open mesquite thickets north of the roadway northeast of Todd Hill.
>
> Varied Buntings are more numerous during wet years and less abundant during years of little winter/spring precipitation. Another characteristic of U.S. buntings is that the vast majority migrate southward immediately after their breeding season.

There are three additional buntings that I have been privileged to have seen, one in the United States and two in Mexico. The **Blue Bunting** is a very rare Texas visitor to the Lower Rio Grande Valley; I have seen it there only a couple times. The male possesses deep blue and black features, while the female is a dull cinnamon brown. According to Steve Howell and Sophie Webb, its "song is a varied, sweet warble often with 1-2 separate notes at start, and fading away at the end."

The **Rosita's (Rose-bellied) Bunting** is a Mexican endemic found on the Pacific Slope in the south. I mention both birds in my chapter, "The Jalisco-Colima Circle," in *Birder's Mexico*. And Howell and Webb include illustrations of both species on plate 60 in *A Guide to The Birds of Mexico and Northern Central America*. They described the voice of the Rosita Bunting as "a wet *plink* or *plek*. Song a sweet, slightly blurry warble."

I also included my sighting of the Rosita's Bunting along the Rio Cihuatlán near Barra de Navidád in *Birder's Mexico*, thusly: "The floodplain as far as we could see was little more than pastureland with an occasional farm or orchard. But the Rosita's Bunting ...despite the arid conditions there, we recorded a good number...close to the main highway."

I also recorded the **Orange-breasted Bunting** at Acatlàn in Hidalgo, an area along Mexico's Pacific Coast, west of the Isthmus. It, too, is a lovely bird with an unmistakable plumage with a yellow-orange underside and blue-green back. Howell and Webb state that it sings "a sad, sweet warble, often shorter and slower than other *Passerina* buntings." I wrote about my sighting in *Birder's Mexico*, thusly:

> We were extremely disappointed to find only a thin line of spotty and overgrazed deciduous vegetation along a rather large but drying riverbed. The floodplain as far as we could see was little more than pastureland with an occasional farm or orchard. But despite the rather arid conditions there, we recorded a good number of trip birds during the two hours before dark. Two of these, the Orange-breasted Bunting and Ruddy Seedeater, were lifers.

Also along the rather arid Rio Cihuatlán, we discovered a very scrubby second-growth habitat that looked different than anywhere else we had previously seen; we drove down a side road to see what might be present. It was there where we found a male **Red-breasted Chat**. What a beautiful bird it was! And so unlike the common Yellow-breasted Chat of the United States. The male is spectacularly marked with bluish-gray and white, with a brilliant rosy-red breast and crissum. It stayed in view, constantly fanning its tail for several minutes, and then disappeared into the dense vegetation.

The Red-breasted Chat is another of Mexico's endemics, occurring only along the Pacific slope, from northern Sinaloa to Chiapas.

There is little doubt that warblers are some of our most colorful and fascinating birds. They occur throughout North America and utilize a huge variety of habitats, from the highest foliage to the shrubbery at ground level. But warblers are not alone in providing us with amazing colors and behaviors. Orioles also possess those same unique characteristics. One of my favorite orioles is the Altamira oriole, a tropical bird that is found in the U.S. only along our southern border in South Texas.

Altamira Oriole, painting

Although orioles do not possess the bright vermilion plumage of the male Vermilion Flycatcher, they, nevertheless are some of our most colorful birds. And I have been fortunate to have encountered many species in the United States as well as in Mexico. Perhaps the **Altamira Oriole** heads my personal list of Texas orioles. Found only in the Lower Rio Grande Valley, it is spectacular in several ways. Firstly, it is large and bright orange and black; its head, nape and rump are orange, and its face and throat, as well as its back, wings and tail are coal-black. And the wings possess ragged white wing bars; a binocular view of its throat will reveal a small bluish patch below each eye. It is an easy bird to identify.

The Altamira's voice is loud and distinct. Its song is a disjointed series of whistles and other sounds, some flutelike in quality. Both sexes are very vocal and their songs carry a considerable distance. And their call notes are a harsh, fussing *ike ike ike*, often the best indicator of their presence when hidden in the foliage.

However, the best indicator of their presence is their long pendulant nests, two feet or more in length. They usually are suspended from flexible, slender terminal branches, at least ten feet high, over open space. Only the female builds the nest, while the male waits and preens nearby, keeping watch for intruders. Or he may follow his mate while she gathers nesting material, such as various grasses, striped pieces of yucca, and almost anything that can be woven into a strong basket.

The Altamira Oriole is primarily a Mexican species with a range that barely reaches the United States, and its southern range extends southward along the Gulf Coast, includes all of the Yucatan Peninsula, and southward into northern South America. In fact, its name was derived from Altamira, a city in southern Tamaulipas, Mexico, where it was first collected. It also occurs along the southern Pacific Slope to Central America and Nicaragua.

In the U.S., Altamira Orioles are found primarily in riparian habitats, such as those along the lower Rio Grande. But in the tropics, they occur in a wider variety of habitats: humid to semi-humid woodlands, forest edges, hedges, and gardens. In winter and during migration, they often occur in large, mixed-species flocks.

I certainly can attest to the size of mixed wintertime flocks. Writing in *Birder's Mexico*, about birding the Yucatan at Chichén Itzá, I included the following:

> I counted at least sixty black and yellow or gold orioles of five different kinds in the tree. Altamira, Orange, Orchard, Hooded, and Yellow-backed orioles were all feeding on the tree's small and inconspicuous greenish-brown flowers. Individuals and groups of three or four orioles came and went during the thirty or forty minutes that I watched the flowering kapok tree.

Another fascinating characteristic of orioles is their behavior of chasing each other around various habitats. At Big Bend's Rio Grande Village, especially prior to nesting, the four species which nest in the area spend an inordinate amount of their time chasing one another. I have wondered why they spend so much of their energy in pursuit-chasing rather than courtship. Perhaps the chase is just part of their courtship.

Finding a **Black-vented Oriole**, the first for the United States, rates as one of my most important avian discoveries. I included a note about that day in *For All Seasons, A Big Bend Journal*, thusly:

> May 1 (1960). I found the Black-vented Oriole again this morning at Rio Grande Village. It was in the same location and appeared to be associated with the same six orioles I had seen four days earlier. Ty and Julie Hotchkiss had also found the oriole. They were camped nearby and were filming a show on the Rio Grande for the National Audubon Society Nature Films; they later toured the country with their production. During the next several days, they took extensive footage with 16mm movie film and numerous slides of the bird. I later used one of those slides in an article I published on this new U.S. oriole in the ornithological journal *The Auk*.
>
> Since it seemed that my Black-vented Oriole was going to stay around, I contacted a few birding friends about the oriole. Word spread in a hurry. During the next five months more than five hundred birders visited Rio Grande Village to see this new bird for the United States. It was banded on July 4, and it remained in the vicinity at least until September 19, about one year after the original date of discovery. In 1970, I found it again at the same locality on April 17; it stayed until at least October 10.

Black-vented Oriole

Later, in writing *A Field Guide to Birds of the Big Bend*, I included additional details about my Black-vented Oriole, thusly:

> On that first morning, I watched it for several minutes among the foliage of a thicket along the nature trail. Later in the day I identified it from reading descriptions by Emmet Blake (1953) and George M. Sutton (1951), but I could not find it the following morning or on several follow-up visits to the area.
>
> On April 28, 1969, I again saw an adult Black-vented Oriole less than 300 feet from the location of the first sighting. For more than 40 minutes I watched it and six other orioles (adult female and immature Hooded Oriole, and two females, one adult, and one immature Orchard Orioles) chase each other from tree to tree within the campground. The Black-vented Oriole appeared to be in close association with the immature male Hooded Oriole, which nicely fit Sutton's description of *Icterus wagleri*. Although I have since learned that the species is monomorphic (having a single-color pattern), I was incorrect in assumed that I was observing a possible nesting pair.
>
> These sightings represent the first authenticated records of the Black-vented Oriole for the United States, although there is a questionable sighting by Herbert Brown from the Patagonia Mountains of Arizona in 1910 (Phillips 1968). South of the border, it occurs "from Sonora, Chihuahua, and Nuevo Leon, south through Guatemala and Honduras to El Salvador in winter and northern Nicaragua" (Freidmann, Griscom and Moore 1957). Mexican breeding records nearest BBNP are from 15 miles south of Gomez Farías, Coahuila, where Charles Ely (1962) studied the avifauna in the southeastern part of the state. Gomez Farías is approximately 350 miles from BBNP.
>
> By mid-May, it was evident that the Black-vented Oriole at RGV was not nesting, and that it was not paired. Its behavior gave no indication that it was defending a territory. Yet by midmorning it would usually disappear into the dense floodplain vegetation and often would not return to the campground portion of its range until late afternoon or evening. By 6:30 a.m. it was always back in the campground with many immature Orchard Orioles or the one or two immature Hooded Orioles that still were present. All these seemed to prefer the fruits of the squaw bush, which was ripening throughout May and June. On May 19, I watched *I. wagleri* fed on flowers of desert willow for several minutes, and on June 28 it caught a

cicada, tore the wings off, and consumer the softer parts of the body, dropping the rest to the ground.

In order to obtain close-up photographs for racial identification, as well as to band the bird so that it could be recognized if it returned, I made several attempts to net it between June 28 and July 4. On July 1, I placed a mounted Great Horned Owl, a species that occurs commonly in the immediate vicinity, on the ground near the mist net. *I. wagleri* perched ten feet from the stuffed bird and watched while a pair of Northern Mockingbirds launched attack after attack on the owl until both were caught in the netting. I even drew a Black-vented Oriole on cardboard, colored it with the proper colors, and mounted the drawing on a stick next to the net. This, too, was a failure – *I. wagleri's* only reaction was one of vague curiosity.

Yet it did show interest in people on a number of occasions. Several times I observed it watching campers going about their routine duties, and on one occasion it flew into a tree above two children who were rolling a red rubber ball around on the ground. It sat there watching this activity for about four minutes before flying off to another perch. On only two occasions did I observe it showing any aggression toward another bird, and then only very short chases (15-20 ft.) of female Orchard Orioles. Although *I. wagleri* usually could be detected by the very low, rasping call it gave, like that of a Yellow-breasted Chat or a Scott's Oriole, a song was never heard.

Finally, by moving the nets each time *I. wagleri* changed positions, I succeeded in capturing it on July 4. Closer examination showed that it was in nonbreeding status; it clearly lacked evidence of a brood patch and had no cloacal protuberance. Close examination of the bill and cere showed no indication that the bird had been caged at any time. Close-up photographs of the chest were sent to Allan Phillips, who identified the bird racially as the *wagleri* from eastern Mexico. The chest had a light chestnut tinge.

Black-vented Oriole with band

After carefully photographing the major features of the bird, I placed a band (no. 632-25253) on its right leg (see photo above) and released it. It immediately flew south to the floodplain portion of its territory, dove into the dense vegetation, and was not seen the rest of the day. By July 10, however, it was right back to the same habits and allowed good binocular examination the first half of the morning.

In early August, it became quite shy and had to be searched for among the dense foliage along the nature trail. I last saw it that year on September 19, exactly one year after the original date of discovery. In 1970, I found it again in the same location from April 17 through September 21 and again on October 10. A good number of birders observed the same banded bird throughout the summer, but it has not been reported since.

All during the years that I worked in Big Bend National Park, one of my favorite birds in the lowlands was the **Scott's Oriole.** Hiking across the desert, even in summer, was made special by the high, rich whistle songs of Scott's Orioles. Even if I did not see one, their songs informed me of their presence. When I did get a glimpse of a bird, it usually was in flight from one Torrey yucca to another. Males are striking, neat and trim birds and well-marked with black and yellow. His coal-black hood, back, and tail contrasts with his deep yellow-colored underparts. A careful observation will also see his whitish wing bars.

Its most significant feature for me, however, is its lovely song that is sung throughout the day. It first reminded me of the flute-like notes of the Western Meadowlark, but it does not end abruptly like that of the meadowlark; it continues its rich whistle-notes for a much longer period. And when it flies to another perch it will almost immediately produce another series of songs.

Scott's Orioles

Elsewhere, I found Scott's Orioles more numerous at Joshua Tree National Park than at Big Bend, probably because of the greater abundance of Joshua trees, readily available nesting sites. The pre-nesting period is a busy time for orioles when they spend so much time chasing one another about. Nests, built by the female, are a rather shallow basket, woven from grass, yucca, and other plant fibers. It is constructed within the protection of the very sharp yucca leaves, although they occasionally are concealed in clumps of mistletoe. Both parents feed the nestlings by regurgitating prey, mostly insects, but also some fruit, berries, spiders, and other arthropods, into the begging mouth of their nestlings.

Wherever Scott's Orioles are present, they seem to me to be an intricate member of that avian community. At Organ Pipe Cactus, biologist Kathleen

Groschupf conducted a breeding bird survey in the saguaro cactus community on May 31, 1991. She reported thirty-four species:

> The dozen most numerous birds in descending order of abundance were White-winged Dove, Verdin, Gila Woodpecker, House Finch, Ash-throated flycatcher, Cactus Wren, Curve-billed Thrasher, Gambel's Quail, Black-tailed Gnatcatcher, Gilded Flicker, Mourning Dove, and Brown-headed Cowbird. Several Lucy's Warblers, Black-throated Sparrows, Phainopeplas, Lesser Nighthawks, Purple Martins, Scott's Orioles, and Northern Cardinals were recorded. Lower numbers of Black and Turkey Vultures, Red-tailed Hawks, Costa's Hummingbirds, Ladder-backed Woodpeckers, Bell's Vireos, Violet-green Swallows (transient), Common Ravens, and Northern Mockingbirds were found. And single individuals of the Harris's Hawk, American Kestrel, Brown-crested Flycatcher, and Canyon Wren were also detected. These numbers provide a good idea of the relative abundance of the breeding desert bird life.

Another oriole, but one that is less common wherever I have recorded it, is the **Hooded Oriole.** It is a smaller and trimer species which I also got to know at Big Bend. In *A Field Guide to Birds of The Big Bend*, I wrote that it is an "uncommon summer resident; rare migrant."

> It has been recorded from March 16 and throughout the summer to October 1. Adult males arrive on their breeding grounds from mid-March to early April; it is several days before females and subadults put in their appearance. They are very gregarious at first and often can be found chasing other orioles. Nest-building begins in mid-May; young are fledged in June and July; adults were found feeding nestlings on May 24, 1969; two fledged birds were seen at RGV on May 11, 1970...and a nest containing two young Bronzed Cowbirds was discovered there in a tamarisk tree on July 12, 1970. Apparently, there also is some late nesting; I found a nest under construction near the top of a tall cottonwood at RGV on July 18, 1970.

Hooded Oriole by Larry Ditto

Hooded Orioles are found primarily in the American Southwest, northward along the Pacific Slope to northern California, and south into Northern Baja. Wintering birds move south into Southern Baja and southward along the Gulf Coast to southern Mexico. Rylander described their feeding behavior thusly:

> They slowly and deliberately glean trees and shrubs for insects, less often for berries and fruits; in towns they frequently visit hummingbird feeders as well as flowers. Sometimes they pierce the base of the flower for nectar. When picking caterpillars from beneath leaves, they may hang upside down like chickadees. Although they do much feeding in low growth, they rarely come to the ground.

The other lowland oriole that nests within Big Bend is the **Orchard Oriole**, a summer resident at cottonwood groves, such as those at Rio Grande Village and Cottonwood Campground. Like its cousins, it is very gregarious and spends the early part of its breeding season chasing one another around their territories. Nesting occurs during May, June, and July, although I found an adult Orchard Oriole feeding a Bronzed Cowbird there on July 28.

The voice of the Orchard Oriole also has the flute-like quality of all the orioles, and Rylander added the following:

> Song: a disjointed series of whistles and other sounds, some flutelike in quality. It has been compared to the Baltimore Oriole's song. Both males and females are very vocal, and their song carries a considerable distance. Call: a harsh, fussing *ike ike ike*, often the best indicator of their presence, as they typically stay hidden in foliage...Their flight is generally quick and jerky and rarely prolonged.

Rylander provided another interesting statement about their behavior: "They rarely form groups larger than the family. These are shy, restless birds that generally stay hidden in the foliage. They fly with powerful wingbeats and usually follow a direct course."

Orchard Oriole by Greg Lasley

In *A Field Guide to Birds of the Big Bend*, I added the following details about Orchard Orioles:

> There is considerable post-nesting wandering, particularly among immature birds, and they may be found to 4,500 feet. One was seen along the lower part of the Window Trail on July 9, 1962, and they are regular at PJ and adjacent arroyos from late July to early September. Females and immature birds were banded at PJ on July 21, 1970; August 6 and 20, 1967; September 7, 1967; and September 20, 1969. By early September this bird becomes rare on its breeding grounds; I have seen only one adult male later than the first of September, although adult females and immatures usually can be found with some effort.

Orchard Oriole males look very different than all the other North American orioles. Males possess black and rusty-orange plumage; the entire hood, back, and tail are black and the underside, including the legs and crissum, are rusty-orange.

The **Audubon's Oriole** is not as well-known as the other orioles, undoubtedly because its U.S. range is restricted to South Texas. It is a bird of the yucca grasslands and Tamaulipas scrub habitats that occur in arid landscapes. During the years that I lived in Victoria, I conducted yearly Breeding Bird Surveys in Goliad, Live Oak, Refugio, and San Patricio counties, south of Victoria. I found that Audubon's Orioles were a fairly common bird on my transects, even though they were not always observed; their sweet melancholy songs revealed their presence and abundance.

Rylander again wrote about it song as "several slow, sweet, melancholy whistles, each on a different pitch, like a young boy idly whistling. It is sometimes rendered as *peut poy it*. This species is much less vocal than the Altamira Oriole."

Audubon's Oriole by Greg Lasley

Audubon's Orioles build their nest on yuccas and various other thorny shrubs. It is a small hanging pouch that is woven from grasses, primarily from fresh green grasses, and usually five to fifteen feet above the ground.

In *Nesting Birds of the Tropical Frontier, The Lower Rio Grande Valley of Texas,* Timothy Brush provided an additional comment about Audubon's Orioles:

> Formerly known as the Black-headed Oriole, Audubon's Oriole was named by S. P. Giraud for his friend John James Audubon, who visited Texas but probably never saw the species in life. Audubon's Oriole is attractively patterned in black and yellow with white trim, but it is usually the song that attracts attention to the bird. George Sutton wrote: "I was completely fooled by whistles that I thought were those of a small boy on his way to a fishing hole…pleasing enough to the ear, but a bit 'off-key' and without definite pattern."

Two other North American orioles, both colorful birds that look somewhat alike, are the Bullock's Oriole of the western half of the country and the Baltimore Oriole that inhabits the eastern states and a good part of north-central Canada. I am familiar with both; I even banded both as they migrated up the Virgin River in the southwestern corner of Utah. While at Zion National Park, I operated two banding stations, one just behind my house, located at the base of the Temple of the Virgins near the park entrance, and another just below the park's west entrance along the Virgin River in the little town of Springdale.

Bullock's Oriole, pair at nest, painting

Bullock's Orioles were netted and banded each spring as they migrated along the riverway, no doubt in route to their nesting grounds within the broadleaf vegetation in the park's lowlands. Zion Canyon, with its extensive riverine zone, offered all they needed. In *Birds of Zion National Park and Vicinity*, I added further details:

> Nest-building normally begins soon after it arrives, and Gifford [local birder] recorded territorial pairs on three breeding bird census plots: two pairs on a 160-acre plot in Coalpits Wash from August 15 to July 2, 1979; six pairs on a 60-acre plot at Grafton from May 13 to July 15, 1978; and

two pairs on a 26-acre plot at Springdale from April 23 to July 10, 1978. Post-nesting birds often join other family groups, and flocks may consist of four or five families...They usually remain in the area at least to late July and there are a few additional scattered reports to September 30.

The Bullock's Oriole is distinguished by the male's black crown, orange cheeks, underparts, and rump and large white wing patches. Bullock's Orioles sing a slow song with rich, whistled notes interrupted with guttural notes and rattles. Rylander wrote that its song is a "melodic series of clear, flutelike whistles. The vocalizations of the Baltimore and Bullock's Orioles differ in structure and sound quality, the song of Bullock's usually being regarded as less musical."

Their nests are easily identified because of their pendant character; they normally are hidden among the foliage. Typically, a Bullock's Oriole nest is attached to twigs near the end of a branch, ten to forty feet above the ground. It consists of an oval-shaped bag approximately six inches deep woven of vegetable fibers, inner bark, and horsehair. The nest is usually lined with wood, down, hair, and mosses.

I recall a visit to Lava Beds National Monument in California where I was surprised by the wide diversity of birds for such a baren landscape. True to its name, Lava Beds is dominated by extensive lava flows and scattered cinder cones, lava tubes, spatter cones, and chimneys. The park also contains a variety of plant communities: big sagebrush-grasslands in the north; brushlands, dominated by mountain mahogany, bitterbrush, and waxy current; and pine forests, predominately ponderosa pine, with scattered areas of western juniper and white fir, and an undergrowth dominated by mountain mahogany, and manzanita on the higher slopes. Approximately sixty percent of the 46,500-acre monument is designated wilderness.

I found a surprisingly large number of birds one morning while walking the Captain Jacks Stronghold Trail. The parking area, surrounded by juniper trees, produced four nesting species: American Robin, Brewer's Blackbird, Bullock's Oriole, and House Finch. There were at least eight Bullock's Orioles near the parking area, chasing one another about the junipers or along the adjacent hillside. And three males sang their guttural, flutelike notes from the junipers or nearby elderberry shrubs. This species was lumped with the eastern Baltimore Oriole for several years, and called "Northern Oriole." It has since been split once again.

Gary Clark wrote about similarities between Bullocks and Baltimore Orioles, thusly:

> The Bullock's oriole certainly looks like a cousin of the Baltimore oriole, but it is distinguished by a black cap rather than a black hood on its head and a distinct black line running horizontally through the eye. What immediately grabs my attention is a big white wing patch like a smear of white paint on the wings, as opposed to the Baltimore's more delicate ribbon-like wing bar.

Baltimore Oriole

Baltimore Orioles are only occasional transients in the western states, and my encounters have been few and far between. Although I have seen these brightly-colored birds several times while birding in the eastern United States, my only photograph of a male Baltimore Oriole was taken in the backyard of my house in Victoria, Texas. The photo above was the result.

I also recall finding Baltimore Orioles while birding the Lakeview Trail at Elk Island National Park in Alberta, Canada. And I later wrote about that day in *The Visitor's Guide to the Birds of the Central National Parks,* thusly:

Waterfowl, as well as land birds, were abundant along the 2-mile Lakeview Trail one June morning. The calls of Ring-billed Gulls along the lakeshore dominated the bird sounds. From the parking area, songs and calls of American Crows, Blue Jays, Red-eyed Vireos, Yellow Warblers, and Baltimore Orioles also were obvious. The loudest were the Baltimore Orioles, singing a rich, sometimes described as a musical, but irregular, series of "hew-li" and other notes. Male Baltimore Orioles are beautiful birds with an all-black hood and upper back, black-and-white wings, and bright orange everywhere else. Females are mostly yellowish with blackish backs and wings. This bird is undoubtedly one of the best-known birds in North America; illustrations are commonplace, used to decorate items from plates to coasters to calendars.

Tanagers are members of the Thraupidae Family, the second-largest family of birds, containing 240 species worldwide. They represent four percent of all avian species. Most are brightly colored fruit-eating birds.

Just as the Bullock's and Baltimore Orioles are representatives of the western and eastern United States, respectively, Western and Scarlet Tanagers fill a similar role. The **Western Tanager** is a well-marked bird with a red face, black back, wings, rump, and tail, and its neck and entire underparts are bright yellow. It also shows broad white wing bars. I have known this lovely bird for most of my life, having grown up in Idaho and Wyoming, where it spends its breeding season in the montane forests.

Western Tanager by Betty Wauer

One of my encounters with Western Tanagers occurred at Lassen National Park in northern California; I remember that day very well. While birding around the Manzanita Lake area, I discovered that Western Tanagers were common in the adjacent pine-fir forests, and I watched them forage for insects among the lichen-covered branches and about the shrubs that bordered the lake. And they often sat high on the tall surrounding trees and sang a song that sounded somewhat like a Robin but hoarser and seldom containing more than four or five phases. Ralph Hoffmann, in *Birds of the Pacific States*, describes its song best: "It is made up of short phases with rising and falling inflections 'pr-ri pir-ri, pee-wi pir-ri pee-wi." They also have a distinct "pritt" or "pri-ti-tick" call. Some birders claim that the song suggests a mood of wildness and freedom.

Perhaps, it was when working at Zion National Park where I got best acquainted with Western Tanagers. They were a reasonably common summer residents and common migrants in the park. In *Birds of Zion National Park and Vicinity*, I wrote that:

> They nest at all elevations, and post-nesting birds have a tendency to wander throughout the high country, and a few move into the lower canyons by early August; numbers increase substantially through mid-September, after which there are only scattered reports until October 10.

A local birder once complained to me that in early July, "Western Tanagers attacked his hummingbird feeders and stripped an early peach tree (along with orioles) in Springdale."

In late fall, most Western Tanagers migrate southward to Mexico and Central America, although a few do remain in the U.S. I recall finding Western Tanagers in the Sierra de Autlán in Colima. They were residing in habitats very similar to those utilized in the states. And in *Birder's Mexico*, I included the following paragraph:

> We also recorded several tanagers within the oak forest, including a pair of Blue-hooded Euphonias and the Red-headed, Hepatic, Western, and Flame-colored tanagers. Rusty Sparrows were fairly common within this habitat, as well. And I found a pair of Black-headed Siskins feeding young in a nest about twelve feet up an extremely thin oak in the center of the trail. It seemed like a poor location for a nest, but three or four youngsters

looked healthy and active, so the parents had evidently chosen an adequate site after all.

The male **Scarlet Tanager** is a real stunner with its scarlet body and contrasting coal-black wings. Females are olive-green with dark gray wings. One of my earliest sightings of this bird was in Ohio at Cuyahoga Valley National Recreation Area. One morning I walked the Ledges Trail that begins at the Happy Valley Nature Center and passes through a forest that seemed to be a good example of the widespread eastern deciduous forest.

The woods were full of birdsongs that morning. Wood Thrush songs were the loudest, and I was totally enjoying their marvelous "ee-o-lay" renditions when a Scarlet Tanager sang its song high in the canopy. It took me several minutes to locate this exquisite bird. Its song reminded me of a Robin's song, but it was not so varied. Aretas Saunders, in *A Guide to Bird Songs*, wrote that its song "is delivered in a rather hoarse, high-pitched whistle. It consists of eight to twelve notes or slurs with slight pauses between them, each on a different pitch from the preceding one."

On another occasion, while walking the Cumberland Trail at Cumberland Gap in Kentucky, I had a totally unexpected encountered with a Scarlet Tanager. I wrote about that incident in *The Visitor's Guide to the Birds of the Eastern National Parks*, thusly:

> A sudden flash of bright red and black in the upland forest is likely to be a male Scarlet Tanager. Its scarlet body and coal black wings are unmistakable. And its loud and cheery song, somewhat like that of an American Robin, is a rather short but melodic "quer-it, queer, query, querit, queer." It will often sing repeatedly from the very top of a high tree. The female is an overall olive-green color, better camouflaged to attend the nest and fledglings while the brightly colored male is defending the territory by singing and, if necessary, attacking another male that may venture too close. This bird is one of the park's most common species, especially at higher elevations.
>
> It took me several minutes to find this songster in the high canopy, but my search was worth the time and effort. The Scarlet Tanager is one of our most beautiful songbirds. The contrast of its scarlet body with its black wings and tail and its silvery bill makes it a wonder of the bird world. And

> seeing a male Scarlet Tanager among the forest greenery that day is one of my most cherished memories.

In *The Birder's Handbook*, Paul Ehrlich and colleagues provided an interesting comment about the relationship between Scarlet and Summer Tanagers:

> Where range overlaps with Summer Tanager, the two species respond aggressively to each other's songs and countersing; coexist by partial habitat shift maintained by interspecific aggression. Male occ feeds incubating female. Female broods for ca. 3 days, and in cold and rain. Females tend to forage higher than males; females forage by hawking far more than do males.

I have been fortunate to have enjoyed two other tanagers in North America: Summer Tanager and Hepatic Tanager. **Summer Tanagers** are widespread in the southern states, utilizing a variety of habitats, from deciduous woodlands to riparian areas. I am most acquainted with this all-red tanager from my years at Big Bend National Park where it resided in cottonwood groves along the Rio Grande as well as in shady woodlands at slightly higher elevations.

Summer Tanager by Greg Lasley

Summer Tanagers are gregarious birds that often occur in the same habitats with Orchard and Hooded Orioles. They may even participate is their mutual behavior of chasing one another prior to nesting. Singing birds are rather persistent prior to nesting and their songs are rather sharp and composed of three or four clear "sweet" syllables, followed by a melodic buzz and a drawn-out trill. Nesting gets underway by early May, and young are out of the nest in June, although there is some late nesting as well. There is a noticeable decline in Summer Tanagers by the last of September, but some years a few may remain over winter.

I also found Summer Tanagers nesting within the riparian vegetation along Beaver Creek at Montezuma Castle National Monument in Arizona. I wrote the following:

> Although it is usually less conspicuous than the Western Kingbird [also resident there in summer], the male Summer Tanager is difficult to ignore when it does come close enough to be seen. During courtship it often dashes about the lower foliage in pursuit of its mate…this seven-inch bird sings a song that sounds at first like that of an American Robin, although it is more hurried and possesses a slight trill. During the breeding season this tanager sings all through the day, but it rarely sings after nesting. Then it can be located by its rather dry "kit-it-up" call notes.

Summer Tanagers reside in many of our parks and refuges throughout the American Southwest and the Southeast. For one example, I recall a day birding near Sandstone Falls at New River Gorge National River in West Virginia. Cliff and Barn Swallows were flycatching over the water and adjacent fields. I parked in the roadside parking lot and walked the boardwalk trail to the edge of a little island where there were several good views of Sandstone Falls. Several songbirds were present among the sycamores and willows. I wrote that:

> The loveliest bird of the day was a bright red male Summer Tanager. It appeared from the foliage of a spreading sycamore tree, flying up after an insect that it captured on the wing, then flying back to an open perch, where it commenced to devour its rather large prey. I have long admired this beautiful creature, which prefers the broadleaf habitats near wetlands. It seems to me that its black-winged cousin, the Scarlet Tanager

> of the upland forests, gets more attention, but I prefer the all-red Summer Tanager.

The other North American tanager – the **Hepatic Tanager** – receives even less notice, mainly because it spends its summer months in rather out of the way places in the Southwestern mountains. Like the Summer Tanager, it also is a red color, but a liver-red color, unlike its bright red cousin. Also unlike the preferred habitats of the Summer Tanager, the Hepatic Tanager spends it breeding season in pinyon-juniper woodlands and drier forests of the Southwestern mountains.

Hepatic Tanager, pair at nest, painting

I got acquainted with the Hepatic Tanager while working at Big Bend, where it is an uncommon summer resident and migrant. I summarized its status in *A Field Guide to the Birds of the Big Bend*, thusly:

> Records extend from April 12, though the summer to October 19. Although this liver-red (male) bird is never numerous, it occurs regularly at a few localities. Breeding pairs usually can be found in upper Green Gulch, above the Chisos Basin cottages, at Juniper Flat, and above Laguna Meadow almost any morning in May and June. I found a nest on a pinyon pine near the Chisos Mountain Lodge on May 12, 1968, and saw an adult male carrying food above the cottages on June 12, 1967.

On a few occasions while watching this tanager above the lodge, I heard its rather distinct call, an abrupt "chuck chuck chuck chuck," and while hiking above the Basin, its loud chuck song seemed to accompany me along my route. At first I was unsure whether its call was a tanager or a Black-headed Grosbeak, a bird that was far more common in the Chisos Mountains than the Hepatic Tanager. But each time I was able to find the singer.

The general behavior of Hepatic Tanagers is somewhat different from other tanagers. When I slowly moved closer to one for better look, I was surprised when it did not immediately fly off; it seemed rather curious about the big hunk of human that was approaching. Then, when I got to within about ten feet from where it was perched, instead of flying off a distance, it flew only a dozen feet away into a pinyon where it seemed to be waiting my next move. I didn't have a camera at the time to try for a close-up photo, but I watched it for another five to ten minutes before it flew away to where I could not see it. I wondered if it had a nearby nest.

Migrants can hardly be separated from early breeding birds and post-nesting wanderers. However, on September 25, 1966, I recorded a "wave of birds near the East Rim at 8:30 a.m." that included at least a dozen Hepatic Tanagers, along with several Ruby-crowned Kinglets, Hermit Warblers, and Vesper Sparrows. The Chisos Mountains apparently are just one the highlands along their migration route.

Wrens are fascinating birds that can be found in every state; from Alaska to Florida. Only the Winter Wren nests in Alaska, but the remainder are spread out throughout North America. I have been fortunate to have recorded them all and in a wide variety of habitats. And of the six species, it is the **Carolina Wren** that is the only resident where I currently live in central Texas.

The world is full of brightly colored birds of all sizes and characteristics. Some, such as Vermilion Flycatchers, are hard to miss. Others lack bright plumages, but their presence, nevertheless, can readily be detected. One of those - the tiny **Winter Wren** - must be searched for in order to see it well. It is a deep-forest imp that too often is heard but not seen.

In my chapter on Pukaskwa National Park in Ontario, Canada, I called it a "mystery bird. Their wonderful songs seem to go on without end, filling the air with the sound of miniature silver bells." Bent reported that a winter wren's 7-second song possesses 108 to 113 separate notes. And Chapman wrote that its song is "full of trills, runs, and grace notes, it is a tinkling, rippling roundelay."

On another occasion, during our post-retirement visits to all the national parks, I spent three days exploring Isle Royale National Park in Lake Michigan. My most memorable happening at Isle Royale was coming face-to-face with a moose as I was walking back to my cabin after dinner. It slowly walked off the trail, seeming to ignore me completely. But I was not so calm; the memory of being treed by a moose many years earlier in the Grand Tetons created concern that took me several minutes to come back to earth.

Isle Royale National Park

I wrote about my Winter Wren encounters on Isle Royale in *The Visitor's Guide to the Birds of the Central National Park*, thusly:

> Of all the common land bird species at Rock Harbor, the Winter Wren may be the most antithesis of the Herring Gull [the most abundant bird at Isle Royale]. This tiny forest bird is seldom seen because of its love for the shadowy undergrowth. However, it is impossible to be out of range of its continual singing, which sounds like distant tinkling bells in the forest. Its song is so constant that the hiker may take it for granted.

Sightings of this tiny songster are often serendipitous; one may suddenly appear at a campsite or while sitting in the woods. They are constantly searching for insects and spiders, creeping in and out of every nook and cranny. But once one of these songbirds is located, it may surprise the birder by its tiny size for such a loud and dominating song. Only an inch in length, it is reddish-brown above, with a pale breast, dark barring on its belly, and a stubby tail.

Winter Wrens nest in dark places under roots and trees, or even in deserted cabins, and the male usually builds one to four dummy nests to fool predators. But it is their tinkling bell songs that are most memorable. They are amazing balls of song.

I am most familiar with the **Bewick's Wren.** It is a bird of the western states with a range which extends from British Columbia to Mexico's Baja Peninsula, east into Texas, and south throughout much of Mexico. It is twice the size of the Winter Wren with a much longer tail, rusty to dark brown upper parts and whitish throat and belly, and a very distinct white eyeline. Also like its smaller cousin, Bewick's Wrens sing a loud, resounding song that seems constant in spring, summer, and fall.

Bewick's Wren by Greg Lasley

At Big Bend National Park, Bewick's Wrens are abundant summer residents and a fairly common migrant and winter resident. It is, in fact, the most numerous summering bird in the park's pinyon-juniper woodlands and only slightly less numerous at thickets down to about 4,000 feet. Migrants move through the park from early March to early April in spring, and from late August to October in fall. Migrants and wintering birds can be expected anywhere, but from April 10 to August 27, it occurs only above approximately 4,500 feet. Bewick's Wrens are fascinating birds that can be found in every state; from Alaska to Florida. Only the Winter Wren nests in Alaska, but the remainder are spread out throughout North America. I have been fortunate to have recorded them all and in a wide variety of habitats. They seem to reside in a wide range of locations.

I even found this wren in the extremely arid Dead Horse Mountains that make up the eastern edge of the park. I included my brief survey of the Dead Horse in *For All Seasons, A Big Bend Journal*:

March 5 (1971). I spent the entire day on Sue Peaks, the higher part of the Dead Horse Mountains. U.S. Geological Survey scientists were remapping the area by helicopter, and the pilot kindly gave me a lift in the early morning and picked me up late that afternoon. Although my trip started without a hitch, my scheduled pick-up did not go as well. I was at the appointed location at the proper time, but the day had become overcast, and my mirror, which I used to signal my location, was not very effective. The helicopter pilot made three large circles trying to locate me and even followed the ridge for several miles north before turning back and making one last try. At the very last second, one of the crew spotted me.

> I detected only seven bird species during my full day on Sue Peaks: Red-tailed Hawk, White-throated Swift, Bewick's Wren, Spotted and Canyon Towhees, Rufous-crowned Sparrow, and Dark-eyed (Oregon) Junco. Sue Peaks supports only scattered red-berry junipers and a variety of desert shrubs and grasses; small stands of junipers and pinyons occur in sheltered canyons below the crest. There is no spring or standing water anywhere.
>
> I also found several land snails among the high limestone rocks on Sue Peaks. I later sent specimens to Lloyd Pratt at the Dallas Museum of Natural History, who claimed that at least one of the three species, a "liptooth" of the genus *Polygyra*, represented a new species. In an August 1995 telephone conversation with Lloyd (then at the University of Nevada at Las Vegas), he informed me that my new discovery was never written

up. The two additional land snail species I found that day included specimens of the distorted metastoma (*Metastoma troemeri*), a widespread land snail, and the Stockton Plateau three-band (*Humboltiana texana*), that represented the westernmost record.

Although I have heard nothing further about my discovery, I am assuming that sooner or later the world will possess a new liptooth, named *Polygyra waueri*.

I had another experience with land snails in Big Bend. I also included that incident in *For All Seasons, A Big Bend Journal*:

June 9 (1968). Laguna Meadows was bright green from recent rains. Several Big Bend century plants were in full bloom; each provided an important food source for a variety of wildlife. My interest today, however, was targeted at the base of the already dead century plants. I turned over several of these dry, brown skeletons in search of land snails. Snail biologist Lloyd Pratt of the Dallas Natural History Museum was eager to receive specimens of a rounded snail with three brown bands that had not yet been described. I collected four dead snails and a single live specimen during a two-hour search and later presented these to Lloyd for his examination. In 1971, based upon these and other snails found in the Laguna Meadows vicinity, he described a new land snail that he named "agave snail" (*Humboltiana agavophile*). Pratt claimed that the species is found closely associated with Big Bend's century plants and is restricted to the uplifted block of Boquillas limestone at Laguna Meadow.

Agave Snail, Big Bend N.P.

Unlike Bewick's Wrens, that can be found throughout most of the American Southwest, the larger **Cactus Wren** is pretty-well restricted to the desert environment. And it is next to impossible to miss because of its loud and raucous songs and conspicuous football-sized nests. Nest-building occurs year-round, and almost every kind of tree, shrub, and man-made structure have been used. I provided a description of a Cactus Wren nest in *Songbirds of The West*, thusly:

> Cactus Wren nests are flask-shaped structures of grasses, small sticks, strips of bark, and other debris, built among the plant's protective spines and sharp leaves. Although the nests often appear messy and poorly constructed, a second look will reveal a rather intricate pattern. Each nest is fully enclosed and waterproof; the insulated inner chamber, lined with feathers, is reached through a narrow passage built at one end near the top.
>
> They also are known to take advantage of a variety of unexpected structures to nest. Once at Big Bend, my neighbor found a nest in the fold of a sheet, and another time in the pocket of his field pants.

Cactus Wrens are one of the bird world's most fascinating creatures. They often will sit for long periods of time at the very tip of a yucca stalk or other tall structure, surveying their domain, and every now and then sing their unique song: a low, rough "choo-choo-choo" to "chug chu-chug-chug," "cora-cora-cora-cora," and other variations. All sound like a car refusing to start, according to Scott Terrill, in *The Audubon Society Master Guide to Birding*. Then, after proclaiming their territories, they will glide down to their mates and greet one another with peculiar "growls" and posturing, crouching with tails and wings extended.

Cactus Wren

Cactus Wrens may be shy and remain hidden, but at other times they are bold and inquisitive. If pursued, however, it quickly dives into a thicket, bush, or cactus patch, usually spreading its tail as it quickly slips out of sight. If perched, it often jerks its tail at intervals much the same as some flycatchers. Another fascinating characteristic of Cactus Wrens is their habit of being late sleepers on bad-weather days. They may even remain in their waterproof nest all day long.

And they seem to eat almost anything they can secure: beetles, ants, wasps, grasshoppers, bugs, some spiders and an occasional lizard and tree frog; also some fruit - elderberries, cascara berries - some seeds, sometimes visits bird feeders for bread, pieces of raw apple, and fried potatoes. I wrote about a Cactus Wren searching for food at White Sands National Monument:

I watched one individual in the parking lot searching the grills of several newly arriving vehicles for insects. It would fly up and extract a butterfly, grasshopper, or some other insect, and then fly back a few feet to consume its prey. After six or seven insect snacks, it flew to the ground in an adjacent planter and began to sort through the debris. I watched it lift several pieces of yucca leaves and peer underneath for prey. If nothing was found, it left the material undisturbed, but on two occasions it threw the material aside with a quick twist of the head and grabbed up the prey found underneath. It seemed especially adept at this method of hunting.

But my all-time favorite wren is the **Canyon Wren**, a resident of the canyonlands of the Southwest and a unique songster. I got to know it best while working at Zion National Park in Utah and during many trips to Grand Canyon National Park in Arizona.

Canyon Wren

One day in May, standing at Grand Canyon's Yavapai Point, I watched the morning light gradually crawl into the canyon, lighting the layers one by one, until it finally exposed the bright green vegetation fringing the river that had carved this stunning landscape. Phantom Ranch, at the bottom of Bright Angel Canyon, glowed like a distant emerald. Just upriver was the

Kaibab Bridge, and directly below my perch was the threadlike Bright Angel Trail, a connecting link from the South Rim.

The morning was alive with birds. My gaze into the depths was suddenly interrupted by a dozen or more White-throated Swifts careening along the cliffs so close to where I sat that I actually felt air movement. Violet-green Swallows were also there, diving here and there in their pursuit of breakfast. And a pair of Ravens made their presence known by their loud "caw" notes. But of the dozen or so birds present that morning, none provided me with such thrilling songs as the Canyon Wrens.

Grand Canyon, South Rim

I was totally enthralled as I listened to their series of slowly descending whistle-notes which echoed from the canyon depths. The song continued as I slowly walked toward the sound. It took me several minutes to find the little songster that suddenly popped into view; it seemed to appear just for me. I froze in place so as not to frighten it and slowly raised my binoculars into place. What a lovely bird it was! Its snow-white throat and upper breast, rusty belly and back, with numerous black spots, and grayish head were conspicuous. And it's extremely long bill, for so small a bird, seemed almost out of proportion.

It suddenly sang again. I listened as its silvery notes descended down the scale like liquid "tews," ending with a mild "jeet." A second later it disappeared behind a shrub, only to reappear a few seconds later where it searched a deep crack for food, creeping forward like a tiny rodent. Then it flew to a nearby rock with a green caterpillar held tightly in its bill. And immediately on swallowing its prey, it sang again, an even louder and longer version of its long, descending, and decelerating song, "TEW, TEW, TEW, tew, tew, tew, tew."

In *The Behavior of Texas Birds*, Rylander added a comment about its voice: "Song: a series of silvery, bell-like, whistled notes that descend in pitch and end in a buzz. Some listeners hear laughter in this song; others, whimsey. Surely it is one of the most unmistakable and unforgettable of all passerine songs. Call: a low peupp."

Canyon wrens also are the avian world's most expressive part of the high cliffs and secretive niches of Zion National Park. Their marvelous songs can be expected everywhere. During May 1963, a pair built a nest in the stone wall of the Zion Inn (now the Nature Center); visitors watched the adults feeding young along the walls and rafters. I also found five fledglings in the Court of the Patriarchs on June 19, 1963, and another nest on the switchbacks below Scout's Landing on June 15, 1965

Rylander also reported that in the 1950s, one pair (possibly more) nested in the dome of the State Capitol in Austin, "entering the building through a broken window. They were able to navigate through corridors, houses, and other human structures."

Zion NP, East Rim

Many of the more elusive songbirds can best be located by their very distinct songs. Vireos provide some of the best examples. Obtaining good, satisfying observations of Bell's, Gray, and Black-capped Vireos usually require time and patience. One of my long-term nemeses to seeing it well is the tiny but loud-mouthed **Bell's Vireo**. *In A Field Guide to the Birds of the Big Bend*, I wrote that it is an "Abundant summer resident (March 8-September 29) along the Rio Grande floodplain." I added the following:

> It also occurs in smaller numbers at other suitable places, such as Dugout Wells, the Old Sam Nail Ranch, Oak Creek, and similar thickets up to 4,500 feet. Early spring arrivals and late fall birds are quiet except for very early morning singing. By March 24 their song is one of the most commonly heard along the floodplain. Nesting occurs from April through June. Most of the summering birds move out of the area by the second week of September, but an occasional individual can usually be found for another couple weeks.

I also recall a Bell's Vireo very early one morning at Amistad National Recreation Area, located along the Rio Grande very near Del Rio, Texas. The dawn literally exploded with birdsong. Northern Cardinals, Yellow-breasted Chats, and Northern Mockingbirds had sung their territorial songs throughout much of the night. And with the dawn, other nesters joined the choir. The mournful calls of Mourning Doves were detected next. Then Bewick's Wrens and Bell's Vireos added their voices to the concert. Bewick's Wrens sang rather complex songs of variable notes ending with musical trills. Bell's Vireos' songs were distinct because of their characteristic of asking then quickly answering questions. Rylander also provided his interpretation of Bell's Vireo's vocalization:

> Song: an unmusical chatter, suggesting a White-eyed Vireo in quality, but more jumbled; it has been rendered as *keedle keedle kee? Keedle keedle koo*, as though asking and answering a question, then answering it...Call: a harsh, scolding *toh wheeo ski*, given especially when disturbed.

During courtship, Bell's Vireo males will chase the female, leaping and fluttering in front of her, and then following her, spreading his tail and singing. During the early stages of nest building, both males and females posture and display before the other. Their open nests are subject to parasitism from Brown-headed Cowbirds; nesting birds sometimes destroy and carry away the cowbird eggs, but other times the vireos will build a second floor over them.

Gray Vireos reside in some of the most remote places in the Southwest. Their breeding range is divided into three or four areas: the largest is in the south-central portions of Nevada, eastern Arizona, and western New Mexico; there also is an elongated area of southern California that extends into central Northern Baja; an area of southeastern New Mexico; and the Texas Big Bend region.

My knowledge of this plain little songbird was derived from the years I was acquainted with it in the Big Bend area. In *A Field Guide to Birds of the Big Bend,* I wrote that it is a fairly common summer resident at suitable localities; rare migrant and winter resident. I added the following:

> Males arrive on their breeding ground as early as March 17 and can easily be detected by their distinct three-whistle song. Males vigorously defend their territories throughout the breeding period and are one of the park's most vociferous species at that time. Six singing males were found in Blue Creek Canyon on May 8, 1960; for about one mile above the ranch house, territorial birds were found approximately every quarter-mile. A more accessible breeding habitat lies directly across Oak Creek Canyon from the Basin. This is where I photographed a nesting bird in early June (see photo below):

Gray Vireo in nest

Records of migrants are scarce, although I have found an occasional loner in the lower foothills and along the Rio Grande. And winter sightings are even less common; those records are from brushy washes below 3,500 feet. I wrote about one sighting in *For All Seasons, A Big Bend Journal*, thusly:

> December 10 (1969). The Chimneys Trail begins at a small pull-off along the Ross Maxwell Scenic Drive and Mule Ears Overlook. The trail skirts the northern edge of Kit Mountain and passes a series of volcanic chimney-like spires, from which its name was derived, and then continues west to Lunas, on the Maverick-Santa Elena Road.
>
> Temperatures were in the fifties and sixties throughout the morning hours, making it a wonderful day to wander and to study anything that attracted my attention. Black-throated Sparrows were especially numerous on the open creosotebush-dominated flats...Of special interest was the Gray Vireo that sang briefly after I spished loudly to attract birds into view; this was the first winter record for Texas. I later told Dr. Jon Barlow, who was studying vireos in the Southwest, about this sighting, and that led to several additional records of wintering Gray Vireos in the park, and eventually to an article in the *Canadian Journal of Zoology* (Barlow and Wauer, 1971).

The Gray Vireo is a nondescript little gray bird with a long tail. Because of its gray features, it easily blends into its equally bland environment. But on its breeding grounds, the male's constant singing of three whistle-notes at a moderate, high, then lower scale, respectively, are repeated over and over. Rylander wrote that its song is "liquid whistles, recalling the blue-headed vireo's song. It has been rendered as *chee bee, choo bee chick.* Its call is a harsh, scolding *schray.*" He also commented on its feeding behavior:

> When Gray Vireos search for insects, their principal food item, they tend to stay in the upper part of trees that are less than 10 feet tall. They only occasionally come to the ground. They move about jerkily, like a wren, and flick their tail more like a gnatcatcher than a vireo...Their flight is quick and usually from bush to bush. They defend territories in winter.

Unlike the very plain Gray Vireo, the **Black-capped Vireo** male is bright and colorful. Black-caps possess an olive-gray back, a pair of whitish wing bars, pale yellowish underparts, and a glossy-black head with white spectacles and red eyes; a dramatic appearance. Females are similar but lack the contrasting black-and-white facial pattern.

Black-capped Vireo by Greg Lasley

This is another bird in which I got to know in Big Bend's Chisos Mountains where it is a summer resident (April 19-August 19) at a very few areas. One of those special places included Campground Canyon, the steep canyon that lies on the south side of Pulliam Peak and is best accessed from Oak Creek Canyon in the Chisos Basin. I wrote about my first Black-cap sighting in *For All Seasons, A Big Bend Journal*, thusly:

> May 18 (1967). I accompanied Dr. Jon Barlow of Canada's Royal Canadian Ontario Museum, on a search for Black-capped Vireos today. Jon is the world's premier vireo specialist and is studying vireos throughout the Western Hemisphere. We spent the entire morning in the Chisos Basin, along the south-facing slope of Pulliam Peak. We succeeded in locating two pairs of these fascinating little vireos, of which are known to nest only within scrub oak habitats in a broad band from north-central Mexico to central Oklahoma.
>
> Male Black-capped Vireos sing a very distinct song of variable notes of grating, squeaking, and clicking sounds, and they will carry on for long periods of time. We detected one male in Campground Canyon, and we gradually worked our way up the very steep slope to where we had a good view of the obviously territorial bird. It was moving up and down the canyon, singing continuously from various shrubs. Our presence seemed to create minimum disturbance. Jon was able to tape an extended period of song.
>
> The day was extremely warm, and I found a bit of shade under a Gregg ash, from where I was able to watch the singing vireo and also to see a good length of the rather narrow canyon. I suddenly found myself staring at a nest in a Texas buckeye not more than thirty-five feet away. After a careful study of the nest through binoculars, I discovered that it contained a bird that was barely peeking over the top. I edged slowly up the slope and somewhat closer to a point from where I could get a better view into the nest. It contained a Black-capped Vireo female, duller than the male with its black head and white spectacles, but a Black-capped Vireo nonetheless. I called Jon over and we stared together at the nesting vireo. Finally we slowly moved forward, and I actually got to within three or four feet of the nest before the female flew off. A quick look at the nest, which held three eggs, and a few photographs later, we both backed away and sat down to watch. The male, all the while we were near the nest,

scolded us vehemently with loud "tcheee" notes. But as soon as we backed off, he resumed singing. Within another few minutes the female returned to the nest, inspected it for any disturbance, and soon settled in again. She seemed undisturbed, although we could see her reddish eyes watching us for any additional threat we might present.

Campground Canyon, Chisos Basin by Betty Wauer

The **Hutton's Vireo** is also a summer resident in the Chisos Mountains. In *A Field Guide to the Birds of the Big Bend*, I wrote that it is a "Fairly common summer resident and uncommon winter resident within the Chisos woodland." Nesting takes place during April, May, and June; a nest located in a clump of mistletoe on an oak at Laguna Meadow on April 29, 1935, was the earliest report of nesting in the park.

Nesting birds are generally restricted to the upper parts of the woodlands, above 5,800 feet. I counted eleven singing birds between the upper Chisos Basin and Laguna Meadow on March 30, 1968. Its loud *sweeet* call can be heard for more than a hundred feet, and a few loud squeaks or hisses in the proper habitat can usually attract one or two birds. Soon after nesting, it moves into the lower canyons, where it may remain throughout the winter. I found two individuals about three hundred feet below the pinyon-junipers in Blue Creek Canyon on February 1, 1969. I also recorded Hutton's Vireos in Arizona, in the mountain woodlands of Saguaro National Park.

Early mornings in a saguaro forest are a one-of-a-kind adventure that everyone, even people only remotely appreciative of nature, should experience. The tall cactus forest possesses a certain calming effect that we all need in today's technological world.

A principal ingredient of this lowland forest is the birdlife. The avian chorus of birds starts each morning with vigor and excitement among the saguaros; it can be difficult to differentiate species in the clamor. Only after the initial confusion wears off can one begin to identify individuals. This early morning choir continues above the Sonoran landscape into the mid-elevation washes and arroyos. Then, finally the pinyon-juniper woodlands are reached which very possibly can duplicate the avian exuberance at lower elevations. The birder then moves above the excitement of the Cactus Wrens, Mockingbirds, Curve-billed Thrashers, Gambel's Quail, White-winged Doves, and House Finches, to be surrounded by a similar hubbub of the woodlands.

That is where the Hutton's Vireos reside. And although this little imp may not be as obvious as some of its neighbors, such as American Kestrels, Acorn Woodpeckers, Bridled Titmice, and Mexican Jays, its song will nevertheless remind us that it is a significant member of the woodland community. Gary Clark, in *Book of Texas Birds*, described the Hutton's Vireo as:

> A chunky though active, five-inch-long bird that by many have described as looking and acting similar to a Ruby-crowned Kinglet. Its dark olive or olive-gray color topside and yellowish wash underside are distinctive, as are its two white wingbars and white spectacles that don't quite cover the top and bottom of the eyes…Finally, and easily noticed, the Hutton's Vireo doesn't hover-feed like the Ruby-crowned Kinglet.

The overall range of the Hutton's Vireo is rather unusual. It occurs along the West Coast of North America from British Columbia to extreme southern California; it also resides in southern Arizona, southward into Mexico, almost to Belize; it also occurs in the Texas Big Bend Country and southward in Mexico where its range meets the western extension in

Veracruz. Basically, it is found at mid-elevations in the highlands of both the Sierra Occidental and the Sierra Oriental.

California Forest by Betty Wauer

Like many neotropical songbirds that nest in the U.S, most spend their winter months south of the border. I wrote about finding Hutton's Vireos at La Cumbre, Jalisco, in *Birder's Mexico*, thusly:

> On May 1, we arrived at La Cumbre at 5:00 A.M., and immediately hiked up the trail into the humid pine-oak forest habitat. Eared Poorwills called along the lower part of the route until 5:35 A.M. At least three Colima Warblers were singing in the canyon below the trail. The *sweeet* calls of Hutton's Vireos were prominent sounds from the oaks. The canyon wren-like call of the Ivory-billed Woodcreeper was heard several times. And further up the ridge the dominant calls were the *ho-say Marie* of the Greater Pewees.

It was exciting to be part of that Mexican bird community, especially to be within the winter range of Hutton's Vireos.

The tits - chickadees and titmice - are members of the Paridae Family that occurs world-wide. The chickadee name was derived from the bird's distinctive "chick-a-dee-dee" call. The name "titmouse" denotes something small. There are five titmouse species in North America: oak titmouse, juniper titmouse, tufted titmouse, black-crested titmouse, and bridled titmouse. All are full-time residents of pinyon-juniper woodlands.

The Oak Titmouse is a bird of the West Coast; Juniper Titmouse occurs throughout much of the Southwest within the states of Nevada, Arizona, southwestern Colorado, and much of western Arizona; Tufted Titmouse has a huge range generally east of the Mississippi River to the east coast and south to central Florida; Bridled Titmouse occurs only in Arizona, the southwestern corner of Nevada, and southward in Mexico to Veracruz. Black-crested Titmouse is found only in Texas from near the Panhandle to the Big Bend Country and southward through Veracruz.

My favorite is the **Bridled Titmouse**, a fascinating bird that occurs only in southern Arizona; I found it surprisingly common at Montezuma Castle National Monument. I wrote about that day in *Birding the Southwestern National Parks*:

> I followed the self-guided trail one spring morning beyond the visitor center to the "castle," a pre-Columbian cliff dwelling nestled in a shallow cave a hundred feet high on an imposing limestone bluff. The morning was bright and calm, and bird songs permeated the air. The loudest songs that morning was those of House Finches singing from high points...Then less than a dozen feet away, from the low foliage of a hackberry, I was suddenly watching a trim little Bridled Titmouse. Through binoculars, I watched it forage among the new foliage, searching each group of leaves in a rather nervous fashion. The bird seemed extremely shy, and I watched its progress from leaf to leaf.
>
> What a handsome bird it was! Its high crest and black-and-white pattern – reminiscent of a horse's bridle, hence its name – gave it a special appeal. Crossing the white cheeks of the Bridled Titmouse is a bold black line that runs from the bill through and beyond the eyes and then turns downward to the all-black throat. The overall plumage was a buff-gray color, but its back and wings contained a tinge of green.

I also recorded Bridled Titmice in Mexico. After visiting with an old friend, Bill Shaldach in Fortín de las Flores for a couple days, we headed back to Mexico City for a flight home. But our time allowed us to check out a site on Pico de Orizaba for a Slaty Finch that Bill told us about seeing there several years earlier. I wrote about that short adventure in *Birder's Mexico*, thusly:

> We left Fortín early the next morning and drove west through considerable fog for about thirty-five miles to where Highway 150 (toll road) crosses one of the main ridges on Orizaba's southwestern flank. We then left our vehicle along the highway at kilometer post 44 and hiked across country toward the distant snow-capped peak. We followed a ridge for several miles through some very patchy landscape. Abundant and extensive areas of Montezuma pine, smaller patches of jelecote pine with its long drooping needles, cultivated fields of various sizes fenced with rockpiles and huge century plants, and low scrubby areas were all intermixed.

I believe that we located the area where Bill had found the Slaty Finch, but, after two hours of intense searching, we did not find a Slaty Finch. As a substitute, we did find forty-two other bird species. The most interesting of these were Blue-throated Hummingbird, a lone Violet-green Swallow, both Steller's and Gray-breasted Jays, Bridled Titmouse, Hermit and Townsend's Warblers, Collared Towhee, Rufous-capped Brush-finch, and Black-headed Siskin.

I first became acquainted with the **Black-crested Titmouse** at Big Bend National Park. It was then considered only a subspecies of the Tufted Titmouse, but was split from the Tufted Titmouse in the updated *Checklist of North American Birds* in 1957. The North American range of the Tufted Titmouse extends from the eastern half of Texas eastward to the Atlantic Coast, including most of Florida. The Black-crested Titmouse occurs only in Texas and in north-central Mexico.

Black-crested Titmouse

Black-cresteds are full-time residents in the Chisos woodlands, generally within the pinyon-juniper woodlands, although they also utilize the cooler canyons, such as Boot Canyon, that are dominated by Arizona pine, Douglas fir, and maple. And they sometimes move into lower elevations in winter; I have even found individuals along the Rio Grande a few times.

Harry C. Oberholser, in *The Bird Life of Texas*, described its "haunts and habits," thusly:

> The Black-crested Titmouse inhabits timber and mesquite along streams as well as *Sabal* palms, live oak mottes of semi-open country, and juniper-oak ravines, gullies, and canyons among hills. The bird also comes about dwellings in country and city, particularly if oak trees or tall shrubbery are present. During the breeding season it moves singly and in pairs, but in winter it occurs in small companies, roaming often with other small tree-dwelling birds.

Oberholser also described its behavior:

> Quick and jerky as it flits amid brush and trees; when moving further, it is still erratic, in the air. Active and inquisitive, the Black-crested works from the very tops of tree to brush, sometimes even on the ground, in search of insects, a few nuts, and occasionally wild fruit and berries. While

foraging, the bird hammers and pecks on the bark to dislodge prey; sometimes it will brace a pecan on a branch and crack it open much like a woodpecker.

Wherever hiking in the Chisos, there seldom was a time when I could not hear the calls of two resident birds: Bewick's Wrens and Black-crested Titmice. The Bewick's Wren's songs were loud and continuous, a "variable, high, thin buzz and warble," according to "*National Geographic Field Guide to the Birds of North America*. The calls of the Black-crested Titmouse are "louder, sharper than Tufted." And Rylander stated that its voice is "A loud, ringing *peter peter peter*, sung persistently throughout the year; also a harsh *day day day* sometimes described as *peevish*."

The tiny **Bushtit** is usually closely associated with the Black-crested Titmouse, utilizing similar habitats, and often joining mixed flocks, primarily in winter. At Big Bend, Bushtits are common full-time residents above the lower edge of the pinyon-juniper woodlands. And they hardly ever are found without other members of their extended family. It was that relationship that first got me interested in these little birds. I wrote about this rather unusual relationship in *A Field Guide to the Birds of the Big Bend*:

This species includes both the black-eared and the plain phases, which are regarded in earlier publications as two separate species. Studies of the bird in Mexico, Arizona, New Mexico, and the Big Bend area have proven that most of the black-eared birds are merely juvenile males. Some breeding males, however, may possess the black-eared coloration. I found a black-eared bird mounting a plain bird near the Window on March 19, 1967. It is also possible to see black-eared birds feeding plain birds and visa-versa. Adults have been found to lay a second clutch of eggs before the first young have left the family; thus fledged birds may actually help feed their younger brothers and sisters. During most of the year, Bushtits occur in flocks ranging from 8 to 10 birds to groups of 45 and 55 individuals.

Bushtits are almost always easy to locate because they constantly call to one another as they move through the woodlands feeding upon insects. Their calls may be heard from as far away as two hundred feet. Although Bushtits have no song, their call is a high-pitched tsit tsit tsit; also, an alarm note sometime rendered as *sre e e e e e*.

During the nesting season, they may be rather difficult to find because they are usually quiet when incubating or feeding young. Pairs generally sleep in the nest at night, even before the eggs have hatched. On winter nights, groups of individuals huddle together at the roost to keep warm. Nests, have been found from mid-March to early June, usually with difficulty, because they are placed in dense foliage of junipers and pines. The nest itself is a long pensile bag with an entrance at the side, composed of various vegetable fibers, such as grasses, ferns, lichens, and moss, and lined with wool or feathers. I found one nest among mistletoe on a drooping juniper in Boot Canyon on May 8, 1968. As the season progresses, there are fewer black-eared birds, although right after nesting, a flock may consist of half black-eared and half plain birds.

Oberholser and Kincaid added some interesting behavioral facts about Bushtits:

> Flight is weak and fitful and when sustained is noticeable undulating. Mannerisms are as lively and agile as a chickadee's, in fact, a bushtit often appears even more nimble in trees due to the longer tail which offers added support and balance. The species is very gregarious except for a brief period during the nesting season after which individuals reband. Groups of bushtits are quite fearless of people or at least so absorbed in foraging operations that they generally ignore any observers.

Birds of the Chisos Woodlands

The painting above, by Nancy McGowan, includes all the resident birds that can be expected in the Chisos Mountains. Included are Mexican Jay, Hepatic Tanager, Acorn Woodpecker, Blue-gray Gnatcatcher, Bushtits, Canyon Wren, Bewick's Wren, and Spotted Towhee.

The **Spotted Towhee**, also known as Rufous-sided Towhee, can be commonplace throughout the woodlands. It is a permanent resident in the mountains; uncommon migrant and winter resident elsewhere. In *A Field Guide to the Birds of the Big Bend*, I wrote the following:

> Nesting birds are confined to the wooded canyons of the Chisos Mountains and nest during May, June, July, and August. The majority seem to remain on their breeding territories throughout the year; one banded at Boot Spring on January 26 was recaptured there again on May 7. Some fall migrants have not been recorded until the middle of October; the earliest sighting along the river is one at Rio Grande Village on October 18. This bird then becomes regular at localized places throughout the winter. Spring migrants reach the Big Bend area by mid-March, reach a peak during the second week of April, and have been

recorded only once after April 21; one at Rio Grande Village on May 15, 1968 (Wauer).

Although some of Big Bend's Spotted Towhees migrate, there still is a constant population in the park. I wrote about one wintertime sighting at Boot Spring in *For All Seasons, A Big Bend Journal*, thusly:

> January 26 (1968). I spent the entire day banding birds at Boot Spring. Using five mist nets placed at strategic places around the cabin site, I netted, measured, banded, and released a total of seventeen individuals of seven species: Yellow-bellied Sapsucker, Ruby-crowned Kinglet, Tufted (Black-crested) Titmouse, Townsend's Solitaire, Spotted Towhee, Rufous-crowned Sparrow, and Dark-eyed Junco (both Gray-headed and Oregon forms).

Their earlier name - Rufous-sided towhee - was a more appropriate name in my opinion. Their obvious rufous sides contrast with the bird's coal-black head, back, and tail. They also possess a pair of white wing bars that vary somewhat. And its song consists of a few introductory notes followed by a trill. Rylander described its song is "a *chip zur eeeeee,*" and its call "a nasal, inquisitive *shreenk*, which at dusk on a cold winter day comes across as remote and haunting."

The Cardinalidae and Emberizidae Families contain numerous small to medium-sized passerine birds. They all are characterized by a stout conical bill that is adapted for eating seeds and nuts. Many possess colorful plumage, and most are full-time residents and do not migrate. These are huge families that are found worldwide. Examples of Cardinalidae birds include grosbeaks, cardinals, and buntings. Examples of Emberizidae birds include towhees, sparrows, and juncos.

The most elegant of all these, for me, is the **Rose-breasted Grosbeak**. Males truly are remarkable birds! Their features include a heavyset body with a coal-black head and back that contrast with their bright rosy breast, and a white belly and rump. Its conical bill is large and pale. In flight they show a large, white triangular patch near the forward edge of each wing.

My first ever Rose-breasted Grosbeak record was while working in Zion National Park. I had established a banding station just below the park's entrance, at the Springdale Ponds, where in three years I banded more than 100 species of birds; one of my most exciting bird was a Rose-breasted Grosbeak. That bird was not only a personal lifer but a new species for the entire state of Utah as well. I took it out of my mist net and instead of banding and releasing it, I retained it for the park's study collection; the date on the specimen tag reads May 3, 1965. Since then, I have recorded several others; I included all my Rose-breasted Grosbeak records at Zion in *Birds of Zion National Park and Vicinity*.

Rose-breasted Grosbeak by Barry Nichols

Over the years, I have been fortunate to have seen this lovely bird on numerous other occasions. Each time was special, almost like seeing it for the first time. One of those times was along the Blue Ridge Parkway in Virginia. I remember that day very well. And I included that occasion in *The Visitor's Guide to the Birds of the Eastern National Parks*, thusly:

> The trees and shrubs behind the Peaks of Otter Visitor Center gleamed in the early morning sunlight. A few leaves were beginning to show autumn color. Bird sound was abundant. Thousands of southbound birds were passing by, mostly in the upper portion of the vegetation, but others were working the brushy undergrowth, and a few were searching for breakfast on the ground. It was a glorious morning.

The most numerous birds that fall morning were the Scarlet Tanagers. The scarlet and black-winged males and yellowish females and immatures were moving through the canopy, four or five at a time, sometimes stopping to feed and other times continuing right to left in what seemed like a steady stream. Rose-breasted Grosbeaks were common as well, and many were calling to one another in a high-pitched "chink." The male's rosy, actually carmine-red, breast and white belly provided a beautiful contrast to its coal black head and back. But the female's dark brown back and streaked breast provided better camouflage. Buff-chested juveniles were also present. All had very great "gros" beaks.

A recall another sighting of a Rose-breasted Grosbeak at Kejimujik National Park in Nova Scotia, Canada. I was walking the Beech Grove Trail, located behind the visitor's center that passes through a dense stand of eastern hemlocks, where I was able to locate a good number of forest birds, when suddenly a male Rose-breasted Grosbeak appeared in all its gorgeous plumage; its rose breast was unmistakable. While watching, it sang a loud, joyous song, somewhat resembling that of an American Robin. Frank Chapman wrote: "There is an exquisite purity in the joyous carol of the grosbeak; his song tells of all the gladness of a May morning; I have heard few memorable strains of bird-music."

Although grosbeaks do not occur where I currently live in central Texas, I am able to enjoy the **Northern Cardinal**, also a member of the Cardinalidae Family. I don't think it receives the admiration deserved, probably because it is so common, but it, especially the males, truly are worthy. The male's cardinal-red plumage, including their tall crest, set off by their black face and short, conical red bill. And females also are attractive in their own way, with lovely buffy underparts; I include photos of both the male and female below.

Cardinal Male

Cardinal Lady

Having lived most of my youth in the West, I knew of Cardinals only from their pictures that appeared on dozens of items: household goods and outdoor advertisements, and then there was the St. Louis Cardinal baseball club; I was San Francisco Giants fan. I honestly cannot remember when or where I saw my first live Cardinal. They did not occur in my first two national parks - Crater Lake and Death Valley - but I did record them during several of my birding trips around the country.

It was in Big Bend, right after my years in Zion, where I was able to enjoy Cardinals regularly. In *A Field Guide to the Birds of the Big Bend*, I wrote about their status: "Common resident in summer and winter at localized areas along the floodplain (such as the Rio Grande Village Nature Trail), and less numerous elsewhere along the river and at adjacent riparian areas up to 3,500 feet; uncommon migrant." I added the following:

> Breeding birds can also be found some summers at Dugout Wells, Government Spring, and the Old Sam Nail Ranch. Most sightings away from the river and other water areas are probably migrants. Small flocks of ten to fifteen birds were seen moving north near Rio Grande Village on March 21, 1969; a lone male was found at Panther Junction on March 19, 1967; and two females were seen at Government Spring on March 20, 1971 (Wauer). One found at Panther Junction on September 16, stayed there from November 29 to December 13, and I banded it on December 9.

Cardinals also were included on Christmas bird counts; I wrote about one count in *For All Seasons, A Big Bend Journal*:

> December 21 (1966). My first Christmas bird count at Big Bend National Park. Four of us – Sharon and I, John Galley, and Roger Siglin – spent the entire day (from 6:30 a.m. to 6:15 p.m.) censusing birds in the Rio Grande Village area. We tallied a total of 961 individual birds of 73 species. The dozen most common species, in descending order of abundance, included 174 Yellow-rumped (Audubon's) Warblers, 115 Northern Mockingbirds, 77 Pyrrhuloxias, 57 Ruby-crowned Kinglets, 32 Northern Pintails, 29 Northern Rough-winged Swallows, 28 Northern Cardinals, 26 Western Meadowlarks, 20 Killdeers, 19 Swamp Sparrows, and 18 Marsh Wrens, Chipping Sparrows, and House Finches (tied).

Ever since my years of living in deserts, at Death Valley and Big Bend, I have loved those quiet landscapes, far from the hassles of city life and what they represent. I sometimes yearn for the calmness they provide. I remember the many times that I stayed still and let the desert absorb my very existence. And as an active birder during those years, a few birdsongs naturally overtook my concentration. And there was no bird sound so memorable than that of the little **Black-throated Sparrow**. In *A Field Guide to the Birds of the Big Bend*, I included the following:

> Nesting takes place from April to June and again in July, August, and September if the rains produce suitable seed plants. This little sparrow can be found in flocks of five to thirty birds during much of the year and can easily be detected by its tinkling calls. It is somewhat shy during nesting. Youngsters lack the typical black throat of adults. This difference may be confusing, but they are almost always accompanied by adults. Black-throats are the park's most seen desert sparrow.

Black-throated Sparrow

I was able to quantify their abundance by breeding bird surveys. I included the details of one of those surveys in *For All Seasons, A Big Bend Journal*, thusly:

> June 2 (1971). For five consecutive years I conducted a breeding bird survey on a twenty-five-mile route (Hot Springs) inside the park, from Hot Springs Junction, past Panther Junction, and west to Todd Hill. This survey was part of a nationwide program initiated by the U.S. Fish and Wildlife Service. My last survey, taken on this date, was typical of all the others. I recorded a total of 771 individual birds of thirty-three species on 50 three-minute stops along the route. The dozen most numerous birds recorded, in descending order of abundance, included Black-throated Sparrow (219 individuals), Northern Mockingbird (108), Pyrrhuloxia (68), Scaled Quail (44), Ash-throated Flycatcher (38), House Finch (36), Cactus Wren (33), Lesser Nighthawk (32), Verdin (27), Mourning Dove (26), Brown-headed Cowbird (25), and Black-tailed Gnatcatcher (24).

Sparrows are seldom on a birder's most wanted list. Except for a few of the rarest, most unique, and endemic species, they often get overlooked. I, on the other hand, have paid extra attention to sparrows. I have enjoyed identifying them all, even when species such as Brewer's and Clay-colored Sparrows require a second look. That is not the case for the Black-throated Sparrow, of the southwestern deserts, and the **White-crowned Sparrow** that nests in the northern areas of North America. White-crowns can hardly be confused with any of the others. Adults are distinctly marked by their black-and-white crowns, pale breast, and brown back and wings with a pair of bars. And I have found white-crowns always to be especially alert and curious. On many occasions I have attracted one or several into the open with spishing sounds. I wrote about one sighting at Canada's Banff National Park in *The Visitor's Guide to the Birds of the Rocky Mountain National Parks*, thusly:

> The Bow Lake area lies along the Icefields Parkway in northern Banff National Park, 25 miles north of Lake Louise; it provides the visitor with a smorgasbord of subalpine habitats. One morning in June, I walked the Bow Glacier Trail that passes through thickets of pussy willows, skirts the forested lakeshore to the glacier runoff, and then follows the floodplain to where it climbs over a series of moraines to the falls. The willow-dominated wetland was alive with birdsong. White-crowned Sparrows, typical wetland species, were numerous, singing their songs of clear whistles and buzzing trills, sometimes described as plaintive notes followed by "more more cheezies please." Their bold black-and-white head patterns, gray collars and underparts, and mottled backs were easily visible in the bright sunshine. I found three individuals hopping about in the center of the trail, occasionally scratching the ground with both feet while searching for seeds.

White-crowned Sparrow

Another of my favorite national parks is Glacier in Montana. Glacier's highlands are much like those in nearby Banff. In fact, the two units, along with Jasper National Park in Alberta and Kootenay and Yoho National Parks in British Columbia, are members of a comprehensive World Heritage Site, established by UNESCO (United Nations Educational, scientific, and Cultural Organization).

I have spent considerable time in these amazing natural areas, hiking several trails and absorbing all they have to offer. I recall one day along Glacier's Hidden Lake Trail, a boardwalk that crossed the open expanse of Rocky Mountain tundra. Millions of alpine flowers dotted the landscape. Their bright reds, purples, and yellows were in sharp contrast to the velvety-green sedges, lichens, mosses, and algae growing at ground level. I wrote about that day as well:

> The 2-mile Hidden Lake Trail, which starts just behind the Logan Pass Visitor Center, provides easy access to the alpine tundra and the high-country bird life. Almost immediately I detected bird song resounding across the open terrain. The songs of White-crowned Sparrows were most obvious. I located one individual perched atop a subalpine fir. Its boldly marked black-and-white head gleaned in the morning light as it sang a variable song of thin whistle-notes followed by a twittering trill.

White-crowned Sparrows also were found in Zion National Park, Utah. Although the park's lowlands are mostly desert, its highlands are not unlike that in Banff. The spruce-fir forest provides nesting sites for several neotropical birds. The most obvious of Zion's forest-birds include Broad-tailed Hummingbirds, Western Wood-pewees, Violet-green Swallows, Steller's Jays, Mountain Chickadees, Ruby-crowned kinglets, Yellow-rumped Warblers, and Dark-eyed Juncos. This highland area, especially around wetlands, also supports White-crowned Sparrows in season.

White-crowns are neotropical species that move southward or into warmer lowlands for the winter months. They then can be found almost everywhere in the southern U.S. and Mexican lowlands, throughout all the Baja Peninsula, and in all the Mexican states, at least south to Veracruz. I recall finding White-crowns one day in Northern Baja. I wrote about that day in *Birder's Mexico*:

> Across the highway from the oceanside pull-off was a little canyon filled with a covering of chaparral. Wrentits, California Thrashers, Bewick's Wrens, Brown Towhees, and Song Sparrows were common there, just as they were within a similar environment throughout California's lowlands. Nearby, among the abundant, wintering White-crowned Sparrows, we later located another chaparral species, the Golden-crowned Sparrow.

Northern White-crowns may move only into the warmer lowlands for the winter. For one example, I include details of a Christmas count in Zion:

> Christmas Bird Counts have been taken in Zion Canyon for several years and provide the best perspective on the wintering populations. The 1990 Zion National Park count tallied 4,134 individuals of 72 species. The dozen most numerous birds, in descending order of abundance, included Dark-eyed Junco, White-crowned Sparrow, House Finch, American Robin, House Sparrow, Cedar Waxwing, Ring-necked Duck, Bushtit, Mallard, European Starling, Pinyon Jay, and Northern Flicker.

My love of birds has been all consuming much of my life, and the opportunities I have had by working in many of America's finest national parks have provided unusual understanding of the space in nature that belongs to birds. Paul Brooks shared his love of birds when he wrote:

> The bond between birds and man is older than recorded history. Birds have always been an integral part of human culture, a symbol of the affinity between mankind and the rest of the natural world, in religion, in folklore, in magic, in art – from early cave paintings to the albatross that haunted Coleridge's Ancient Mariner. Scientists today recognize them as sure indicators of the health of the environment. And as modern field guides make identification easier, millions of laymen watch them just for the joy of it.

Acknowledgements

Several friends provided support during the production of "Some of My Favorite Things." A few helped in significant ways: LeeAnn Nichols was always available to attend to my numerous computer questions and problems. Bill Nichols was a constant advocate for my progress and always ready to help in innumerable ways. Their moral support was much appreciated!

I also want to acknowledge a few individuals who accompanied me on my various adventures and/or freely assisted in one way or another. They include Jon Barlow, Bob Behrstock, Jim Brock, Jan Dauphin, Mark Elwonger, Eric and Sally Finkelstein, Cheryl Johnson; Greg Lasley, Kent Rylander, Ray Skiles, Betty Wauer, Jim and Lynne Weber, Anse Windham, and Lee Ziegler. In addition, Greg Lasley kindly answered each request for the use of six of his superb photographs (numbers 24, 26, 42, 59, 66, and 67); he is given credit on each caption. Larry Ditto supplied photo number 65. And several of the scenics were provided by my wife, Betty, who passed away prior to publication. All the remaining photos are from my personal collection.

Thanks are also due to two artists for the use of their works: Nancy McGowan for the full-page painting of "Birds of the Chisos Mountains," and John H. Dick for two paintings: Blackpoll and Bay-breasted Warblers and the Imperial Woodpecker.

There are a few other relatives and friends who provided inspiration. They include (alphabetically) Jim Brock, Greg Lasley, Barry and Sharon Nichols, Brent Nichols, Brent Wauer, and Jim and Lynne Weber. Their help is very much appreciated.

References Cited

American Ornithologist' Union. 1998. *Check-list of North American Birds*. 7th ed. Washington, D.C.: American Ornithologist' Union.

Bangs, 0utran. 1925. The history and character of *Vermivora crissalis* (Salvin and Goodman). *Auk* 42: 251-53.

Barlow, Jon C., and Roland H. Wauer. 1971. The Gray Vireo (Vireo vicinior Coues; Aves: Vireoinidae) Wintering in the Big Bend Region, West Texas. *Canadian Journal of Zoology* 49(6):953-955.

Blake, Emmet Reid. 1949. The nest of the Colima warbler in Texas. *Wilson Bull*. 61:65-67.

Brush, Timothy. 2005. *Nesting Birds of a Tropical Frontier*. College Station: Texas A&M Univ. Press.

Clark, Gary. 2016. *Book of Texas Birds*. College Station: Texas A&M Univ. Press.

Ehrlich, Paul R., David S. Dobkin, and Darryl Wheye. 1988. *The Birder's Handbook*. New York: Simon and Shuster, Fireside Book.

Halle, Louis J. 1957. *Spring in Washington*. New York: Harper and Brothers.

Howell, Steve N. G., and Sophie Webb. 1995. *A Guide to the Birds of Mexico and Northern Central America*. Oxford: Oxford Univ. Press.

Lasley, Gregory W., et. al. 1982. Documentation of the Red-faced Warbler (*Cardellina trubrifrons*) in Texas and a Review of its Status in Texas and Adjacent Areas. *Bull. Texas Ornithological Soc*., Vol. 15, Nos 1 & 2:8-14.

Lockwood, Mark W. 2001. *Birds of the Texas Hill Country*. Austin: Univ. Texas Press.

_____ and Brush Freeman. 2004. *Handbook of Texas Birds*. College Station: Texas A&M Univ. Press.

National Geographic Society. 1987. *Field Guide to the Birds of North America*. Washington, D.C.

Oberholser, Harry C., and Edgar R. Kincaid, Jr. 1974. *The Bird Life of Texas*. Austin: Univ. Texas Press.

Peterson, Roger Tory. 1990. *Western Birds*. Houghton-Mifflin, Boston, Massachusetts.

Peterson, Roger Tory, and Edward L. Chalif. 1973. *A Field Guide to Mexican Birds*. Boston: Houghton Mifflin Co.

Peterson, Wayne. 2008. *Audubon Society's Master Guide to Birding*. Boston: Houghton Mifflin Co.

Phillips, Allan R, Joe Marshall, and Gale Monson. 1964. *The Birds of Arizona*. Tucson: Univ. Arizona Press.

Raitt, Ralph J. 1967. Relationship between Black-eared and Plain-eared Forms of Bushtits (Psaltriparus). *The Auk* 84:503-528

Rylander, Kent. 2002. *The Behavior of Texas Birds*. Austin: Univ. Texas Press.

Saunders, A. Aretas. 1951. *A Guide to Bird Songs*. New York: Doubleday

Schaldach, William J., Jr. 1963. *The avifauna of Colima and adjacent Jalisco, Mexico*. Los Angeles, Calif.: Proc. Western Found. Vertebrate Zoology.

Sutton, George Miksch. 1951. *Mexican Birds: First Impressions*. Norman: Univ. Oklahoma Press.

Sykes, Paul. 2008. *Master Guide to Birding*. Boston, Massachusetts: Alfred Knopf.

Terres, John K. 1987. *The Audubon Society Encyclopedia of North American Birds*. New York: Alfred A. Knopf.

Van Tyne, Josselyn. 1929. *The Discovery of the Nest of the Colima Warbler (Vermivora crissalis*). Misc. Publ., Univ. Of Michigan, No. 33.

Vickery, Peter. 2008. Master Guide to Birding. Alfred Knopf, Boston Massachusetts.

Wauer, Roland H. 1967a. Colima Warbler Census in Big Bend's Chisos Mountains. *National Parks Magazine*, Nov., 8-10.

_____ 1970. The Occurrence of the Black-vented Oriole, *Icterus wagleri*, in the United States. *Auk* (4): 361-62.

_____ 1971. Ecological Distribution of Birds of the Chisos Mountains. *Southwestern Naturalist* 16:1-29.

_____ 1973. *Birds of Big Bend National Park and Vicinity.* Austin: Univ. Texas Press.

_____ 1980. *Naturalist's Big Bend.* College Station: A&M Univ. Press.

_____1992. *The Visitor's Guide to the Birds of the Eastern National Parks, United States and Canada.* Santa Fe, NM: John Muir Publications.

_____1993. *The Visitor's Guide to the Birds of the Rocky Mountain National Parks, United States and Canada.* Santa Fe, NM: John Muir Publications.

_____ 1994a. *The Visitor's Guide to the Birds of the Northwestern National Parks, United States and Canada.* Santa Fe, NM: John Muir Publications.

_____ 1994b. *The Visitor's Guide to the Birds of the Central National Parks, United States and Canada.* Santa Fe, NM: John Muir Publications.

_____ 1996a. *A Field Guide to Birds of the Big Bend.* Houston, Texas: Gulf Publ. Co.

_____ 1996b. Flammulated Owl Records Following May Storms in Zion Canyon, Utah. *Condor* 66 (2): 211.

_____ 1996c. *A Birder's West Indies, An Island-by-Island Tour.* Austin: Univ. Texas Press.

_____ 1997a. *Birds of Zion National Park and Vicinity.* Logan: Utah State Univ. Press.

_____1997b. *For All Seasons, A Big Bend Journal.* Austin: Univ. Texas Press.

_____ 1999a. *Birder's Mexico.* College Station: Texas A&M Univ. Press.

_____ 1999b *The American Robin.* Austin: Univ. Texas Press.

_____ 2001. *Naturally…South Texas. Notes from the Coastal Bend.* Austin: Univ. Texas Press.

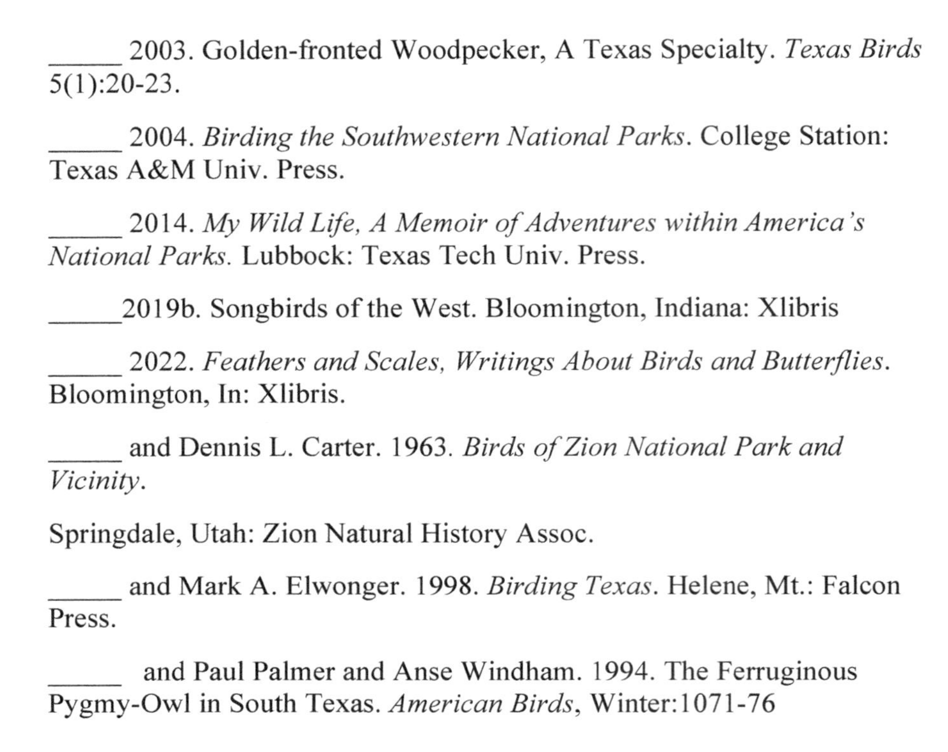

_____ 2003. Golden-fronted Woodpecker, A Texas Specialty. *Texas Birds* 5(1):20-23.

_____ 2004. *Birding the Southwestern National Parks*. College Station: Texas A&M Univ. Press.

_____ 2014. *My Wild Life, A Memoir of Adventures within America's National Parks.* Lubbock: Texas Tech Univ. Press.

_____2019b. Songbirds of the West. Bloomington, Indiana: Xlibris

_____ 2022. *Feathers and Scales, Writings About Birds and Butterflies.* Bloomington, In: Xlibris.

_____ and Dennis L. Carter. 1963. *Birds of Zion National Park and Vicinity.*

Springdale, Utah: Zion Natural History Assoc.

_____ and Mark A. Elwonger. 1998. *Birding Texas*. Helene, Mt.: Falcon Press.

_____ and Paul Palmer and Anse Windham. 1994. The Ferruginous Pygmy-Owl in South Texas. *American Birds*, Winter:1071-76

INDEX

PHOTOS in ***BOLD***

C

D

E

F

G

H

I

J

K

L

M

N

O

P

R

S

T

U

V

W

Z

Printed in the USA
CPSIA information can be obtained
at www.ICGtesting.com
CBHW061239280924
14965CB00059B/71